URBAN MODERNITY

URBAN MODERNITY
CULTURAL INNOVATION IN THE SECOND INDUSTRIAL
REVOLUTION

MIRIAM R. LEVIN, SOPHIE FORGAN, MARTINA HESSLER,
ROBERT H. KARGON, AND MORRIS LOW

THE MIT PRESS
CAMBRIDGE, MASSACHUSETTS
LONDON, ENGLAND

© 2010 Massachusetts Institute of Technology

All rights reserved. No part of this book may be reproduced in any form by any electronic or mechanical means (including photocopying, recording, or information storage and retrieval) without permission in writing from the publisher.

For information about special quantity discounts, please email special_sales@mitpress.mit.edu

This book was set in Engravers Gothic and Bembo by Toppan Best-set Premedia Limited. Printed and bound in the United States of America.

Library of Congress Cataloging-in-Publication Data

Urban modernity : cultural innovation in the Second Industrial Revolution / Miriam R. Levin . . . [et al.].
  p. cm.
Includes bibliographical references and index.
ISBN 978-0-262-01398-7 (hardcover : alk. paper)
1. Urbanization—History. 2. Technological innovations—Economic aspects—History. 3. Industrialization—History. I. Levin, Miriam R.

HT361.U7173 2010
307.76 09—dc22

2009034747

10 9 8 7 6 5 4 3 2 1

CONTENTS

LIST OF FIGURES   VII

PREFACE   IX

1   DYNAMIC TRIAD: CITY, EXPOSITION, AND MUSEUM IN INDUSTRIAL SOCIETY   1
MIRIAM R. LEVIN

2   BRINGING THE FUTURE TO EARTH IN PARIS, 1851–1914   13
MIRIAM R. LEVIN

3   FROM MODERN BABYLON TO WHITE CITY: SCIENCE, TECHNOLOGY, AND URBAN CHANGE IN LONDON, 1870–1914   75
SOPHIE FORGAN

4   THE COUNTERREVOLUTION OF PROGRESS: A CIVIC CULTURE OF MODERNITY IN CHICAGO, 1880–1910   133
ROBERT H. KARGON

5   "DAMNED ALWAYS TO ALTER, BUT NEVER TO BE": BERLIN'S CULTURE OF CHANGE AROUND 1900   167
MARTINA HESSLER

6   PROMOTING SCIENTIFIC AND TECHNOLOGICAL CHANGE IN TOKYO, 1870–1930: MUSEUMS, INDUSTRIAL EXHIBITIONS, AND THE CITY   205
MORRIS LOW

7 CODA   255
MIRIAM R. LEVIN

INDEX   261

FIGURES

2.1 Map of Paris showing streets constructed between 1853 and 1914. 18–19
2.2 Portrait of Adolphe Alphand. 28
2.3 Paris, demolition for the Boulevard Saint-Germain, c. 1870s 35
2.4 Trocadéro Palace, exterior. 40
2.5 Paris, plan of the 1900 Exposition Universelle Internationale. 44
3.1 Portrait of Sir William Henry Preece. 86
3.2 London, map of the principal electrical institutions and undertakings, 1906. 90–91
3.3 London, Wood Lane Power Station. 94
3.4 London, Duke Street electricity transformer station. 95
3.5 Poster for the Japan-British Exhibition, 1910. 106
4.1 Chicago, elevated railway, 1905. 134
4.2 Portrait of Marshall Field. 136
4.3 Chicago, view of the 1893 World's Columbian Exposition. 145
4.4 Chicago, Field Columbian Museum. 146
4.5 Chicago, plan of the complete system of street circulation, 1909. 156
5.1 Portrait of Rudolf Virchow. 171
5.2 Berlin, Museum of Pathology. 182
5.3 Berlin, industrial exhibition in Treptower Park, 1896. 191
6.1 Portrait of Prince Iwakura. 216
6.2 Detail of Toyohara Chikanobu, *Husband and Wife and Beauties on the Sumida River*. 230
6.3 Detail of Utagawa Hiroshige, *Second National Industrial Exhibition at Ueno Park Showing the Art Museum and Fountain*, 1881. 232

6.4   Tokyo, Edobashi area, Shōwa Dōri.   240
6.5   Tokyo, proposed subway map, c. 1921.   241
7.1   Eugène Hénard, the City of the Future.   258
7.2   Mexico City, street scene, c. 1911.   259

PREFACE

Like all books, this one has its history. When I began to consider how to study the emergence of what I then termed a culture of control in the late nineteenth century, I was intrigued by the dynamic running through urban growth, international expositions, and museums that generated the new cultural framework for what we call industrial society. The question was how to make the topic manageable—that is, capable of being completed without a lifetime's research. The answer emerged from discussions I had with Professor Robert Kargon: make it a collaborative and comparative study of selected key cities that could serve as exemplars for the study of other urban locales. The result is the inclusion of chapters on Paris, London, Chicago, Berlin, and Tokyo by individuals knowledgeable about the history of their chosen city.

If the project had stopped there, it might have been just another collection of essays on the common topic of urbanization in the late nineteenth to early twentieth century. We consciously moved outside of this convention by using a set of conceptual and project design constraints that kept us focused on the original idea of the dynamic forces of science and technology that I had originally identified. The conceptual constraint was to set a common framework for approaching the history of each city: each of us looked at the possibility of connections between elites, urban rebuilding, expositions, and museums in our respective cities. It was my role to bring these together through editing and the extended introductory and concluding chapters. As part of our work plan, we held a series of workshops to compare and critique one another's research and drafts over an eighteen-month period. In the end, research and writing went smoothly and rapidly.

It is not possible to list all the people and institutions to which we each owe thanks. A few need to be mentioned here for what we owe them collectively. First, our appreciation to the National Science Foundation for supporting the project. Dr. Ronald Ranger was wonderfully open to the idea of the workshops, as well as to the intellectual concept itself, when I first proposed them to him. Dr. Ranger's successor, Dr. Fredrick Kronz, continued to advise me on negotiating extensions and an application for additional funding. Thanks also to the Department of History of Science and Technology at Johns Hopkins University for hosting and helping fund the Baltimore workshop; and to Dean Cyrus Taylor, the College of Arts and Sciences, the Baker Nord Center for the Humanities, and the History Department at Case Western Reserve University for hosting and helping to fund the Cleveland workshop. My good friend and colleague Professor Catherine Lavenir deserves our gratitude for providing many helpful comments, suggestions, and insights during our discussions. She also is to be thanked for organizing our Paris workshop at the Sorbonne, University of Paris. We also appreciate the warm reception Mme. Hélène Bignon provided us in Montmartre during our Paris stay.

I personally owe a debt of gratitude to Dr. Mark Eddy, Social Sciences Librarian for the Kelvin Smith Library at Case Western Reserve University, for his help in tracking down a number of images and copyrights. My graduate research assistant, James Johnson, worked cheerfully and assiduously to prepare a digitized version of the manuscript, also checking spelling and citations along the way. Michael Berk did a superb job of editing the entire manuscript, helping to strengthen arguments and bring stylistic consistency to the text. As with all publications, misspellings and other mistakes in the text are the responsibility of the authors.

Miriam R. Levin

1 DYNAMIC TRIAD: CITY, EXPOSITION, AND MUSEUM IN INDUSTRIAL SOCIETY

MIRIAM R. LEVIN

At its simplest, modernity is a shorthand term for modern society or industrial civilization. Portrayed in more detail, it is associated with (1) a certain set of attitudes towards the world, the idea of the world as an open transformation by human intervention; (2) a complex of economic institutions, especially industrial production and a market economy; (3) a certain range of political institutions, including the nation-state and mass democracy. Largely as a result of these characteristics, modernity is vastly more dynamic than any previous type of social order. It is a society—more technically, a complex of institutions—which unlike any preceding cultures lives in the future rather than the past.

—Anthony Giddens, from *Conversations with Anthony Giddens: Making Sense of Modernity*

At the end of the nineteenth century, the Lumière brothers turned their new moving picture camera on the contemporary urban scene: Paris, London, Chicago, New York, and even the laggard Moscow, in all their modernist dynamism, were captured and projected onto the screen. Despite regional variations in dress and architectural style, the filmed sequences of the boulevards all show moving crowds and forward-plunging traffic enfolded in an orchestration of wide streets, stone curbs and gutters, sidewalks, streetlamps, telegraph wires, tram tracks, shops, theaters, and apartments. From time to time, there are automobiles. Occasionally, we see people at an exposition, or in front of a museum, exhibit, or public building.

This was the exterior: modernity essentialized into coordinated surface flow. Unseen are the sewers, gas lines, subways, and national and international communication networks that complement the instruments of modern urban life that channel all this human activity in the street above. Also missing from view are the city-based elites: businessmen, industrialists, professionals, and government, administrative, and political leaders whose vision and policies brought this new urban existence into being.

This book is intended to bring these elites and their accomplishments into focus. More than any other historical force, urban elites active in an international exchange of ideas constructed a modern, industrial-based culture by establishing new institutions, programs, and projects related to science and technology. This new culture of change helped tame the social conflict and economic stress arising from industrialization, while creating a human-built continuum of time and space out of the very technologies and scientific ideas that fueled industrialization itself. This revolution took place roughly between 1850 and 1930, a period including what is known as the second industrial revolution. Paris, London, Chicago, Berlin, and Tokyo were key sites, and among the most important centers of action for those who made this revolution.

Urban Modernity examines the ideas and policies embodied in urban planning, international expositions, and museums in these five major urban centers. These cities were at the heart of this historic shift, negotiating between regional and international networks of production, consumption, and exchange. During this period, Paris, London, Chicago, Berlin, and Tokyo underwent similar patterns of industrialization, due to the shared international perspective of their planners, while also exhibiting differences arising from their varied political, social, and economic circumstances. Each participated in a redefinition of time, space, and human social and economic relationships during this period, as small groups of elites sought to shape the characters of their societies and manage industrial growth. All five cities were, in one way or another, marked directly by the coordinated efforts of these groups.

Cultures of science and technology were both grounded in and constitutive of their respective urban cultures. The chapters that follow explore the working hypothesis that rapid economic and technological changes in late-nineteenth-century societies led to conditions of social and political instability. These circumstances demanded new institutions created expressly to manage citizens and take advantage of possibilities for industrial growth. Specifically reacting to class conflict, fear of the unknown consequences of new discoveries, and the weakening of local institutions as nation-states expanded their powers, business and government leaders looked to cities as the loci for organizing new lifestyles, institutions, and professional groups to design and steer the process of modernization. Urbanization marked a significant break with a traditional understanding of society as rooted in agriculture, and required the construction of an entirely new reality in which science and technology would be not only intellectual touchstones but reliable agents of growth. Reformed urban centers, universal expositions, and museums would give birth to this new understanding of existence, restructuring time into a continuous story of positive development, stretching into a relatively risk-free, human-constructed future.

This study examines institutions that characterized the culture of modernity being established in these five cities, along with the contexts and personal networks that explain them. We have chosen as the basis for comparison the following categories:

First, we examine contemporary efforts to build a modern technological urban environment, including both urban construction and development projects and the production of the cadres of professional experts who inaugurated and managed them. We look at discussions among elites concerning the introduction of new technologies into the infrastructures of major metropolises, and examine the founding and reform of the educational institutions, research institutes, government services, and nongovernmental organizations that trained and employed these experts.

Second, we examine the process by which the past was reformulated "scientifically," in conformity with contemporary concerns. Here we look at museums founded during this period, which were created to show that the past confirms the existence of linear progress. We also look at heritage sites that were identified as significant for similar purposes.

Third, each essay considers contemporary conceptions of the future. In the projects studied, the future is generally portrayed as the linear result of scientific and technical progress—safe, increasingly prosperous, congenial, and controllable. In the five cities under discussion, we examine representations of the future human-constructed environment, focusing on the exhibits at universal expositions or world's fairs. These events, in which all of the nations under discussion participated, promoted progress and involved massive coordination centered on urban hubs. We also look at urban design—an important theme in exposition exhibits—as an indicator of what was hoped would lie ahead.

The book begins with an essay on Paris from the 1850s to 1914, focusing on its politicians, administrators, industrialists, social scientists, architects, and engineers, who over the course of two different political systems created a vision of a scientifically administered future and sought to implement it through a congeries of commissions, institutions, agencies, and organizations concerned with urban planning, international expositions, and museums. There are very good reasons for starting with Paris in the 1850s, some of which contemporaries themselves acknowledged. Over the next fifty years Paris emerged as the capital of Western civilization, seemingly poised to exert a strong influence on cities throughout the world in the coming century. This identity was, in part, the product of anxious self-promotion, a reflexive response to competition with and real threats from France's neighbors. In order to modernize, the city undertook large-scale urban development, five international expositions, and the founding of a number of museums. After the revolution of 1848, the city's international status rose dramatically during the 1850s and 1860s. Spurred by a desire to best the English and to control urban unrest, Napoleon III and his prefect Baron Haussmann marshaled authoritarian control over financial and political power to begin to turn Paris into a model of an industrial capitalist city.

When the Franco-Prussian War and the civil war that followed left parts of the city in ruins and the divided population stunned, leaders of the Third Republic renewed their commitment to Enlightenment ideals and liberal democracy, redirecting the cultural agenda of industrial capitalism away from that espoused during the Second Empire. Science and technology provided France's new leaders the means to create a new set of democratized social and economic relations committed to orderly growth. National in scope, but international in its implications, their vision was centered on Paris, and their plans to transform the city focused on secular solutions to the social, economic, and infrastructural problems the city posed. As a result of several factors (advances in science and technology that brought electricity, the airplane, and high-rise construction, as well as the advent of international empire building and shifts in political agendas), by the early 1900s a younger generation faced the challenge of reinventing this culture of change to accommodate a new phase of industrialization.

Meanwhile, from 1870 onward, London had become one of the largest and richest cities in the world, as well as the capital of a vast empire. Concerns with containing and sustaining the urban masses continued to animate the efforts of the city's elites and government in a period characterized by moves toward increasing state centralization and intervention. As London's built area continued to expand rapidly, new transport and communication networks—underground trains, electric trams, and an efficient postal service—were created.[1]

London was the metropolitan center of the largest Western empire, the major locus of its scientific spectacles and knowledge with its museums, exhibitions, schools, colleges, and government institutions. London's parks inspired Napoleon III's plan for Paris, and the first international exposition was held on the city's outskirts in 1851. Yet, unlike Paris, late-nineteenth-century London was itself highly decentralized administratively and allowed much room to individual entrepreneurs. The scientific elites so clearly recognizable in mid-Victorian Britain—the members of the X Club, for example—became more diffuse, separated as they were into numerous specialist associations. Educational institutions and events such as international exhibitions brought together scientific and technological experts in various specialties, in particular electrical engineers such as William Preece, who were involved in creating the modern technological infrastructure of the city.

Even as the city modernized, museum and exhibition culture tended to move out of the central areas of the city to the peripheral suburbs. There, from 1870, the South Kensington complex of science and art institutions, located in a former suburb, provided a contrast to the development of large commercial exhibition venues nearby.

The London chapter of this book focuses first on the physical development of the urban environment, and how cultures of science and technology were both rooted in and constitutive of urban culture. Then, through a look at the city's museums and exhibitions and its educational institutions, it examines shifts in scientific culture from "useful" notions of moral improvement to a broader range of values. Visions of the future were constructed and the historic role of science invoked as the key to a story of progressive improvement, despite some resistance to change, in particular how such changes might interact with the imperial dimension.

Like Paris in the wake of the Commune, Chicago after its devastating fire of 1871 presented a geographic tabula rasa for wealthy urban elites wishing to expand industrial wealth and heal social wounds through urban rebuilding. Before the fire, and especially afterward, Chicago was rapidly built into "nature's metropolis."[2] Between 1880 and the 1890s its population doubled, to over a million souls, making it the second largest U.S. city. This growth was made possible by technical change: railway engineers made Chicago one of the nation's great transport hubs; civil and sanitary engineers built roads, reversed river flows, and erected tall buildings; while inventors and entrepreneurs provided the employment that drove migration from the countryside. With this phenomenal demographic and technical change came class conflict. Chicago was a locus classicus of the growth pangs of industrial capitalism. Not only was civil strife an urgent problem, but popular resistance to new technologies surfaced.

These concerns spawned new civic associations with interlocking directorates, such as the Commercial Club, founded in December 1877. The Commercial Club's avowed purpose was "advancing by social intercourse … the prosperity and growth of the city of Chicago."[3] Over the next three decades, members of this extraordinarily influential club would establish, fund, and often lead scientific, cultural, and educational institutions to ensure the successful propagation of the civic ideal they embodied.

The Chicago portion of this book focuses upon three significant examples of the civic ideal made real, showing how powerful urban leaders forged consensus and promoted what historian Robert Wiebe would call a "revolution in values" regarding scientific and technical change.[4] The three projects discussed are the World's Columbian Exposition of 1893 and Daniel Burnham's Chicago Plan of 1909 that grew out of the exposition; the Armour Institute of Technology, which opened formally in 1893 and soon became the premier producer of the "techno-structure" of Chicago; and the Field Columbian Museum, opened in 1894, which displayed "the progress, the skill and the genius of our race"—a fitting "outcome and the monument" to the exposition.[5]

In contrast to the other two European cities examined in this book, Berlin did not start evolving into a modern industrial and scientific city until the founding of the German *Kaiserreich* in 1871. From then on, Berlin began to lay aside its rather provincial image, and by World War I it had developed into a modern city of science, industry, and culture. It had also become a symbol of the modern metropolis and an international city, a transformation that went hand in hand with the city's evolution as the political center of the powerful new German state.

Berlin is therefore particularly suitable for the investigation of the roots of the modern industrial way of life. The city's population increased dramatically, growing from 800,000 in 1870 to over four million by 1920. Science and science-based industries played key roles in Berlin's development as a metropolis. After the founding of the Reich, Berlin became a center of scientific research. In response to the requirements of industry, urban elites established the Physikalisch-Technische Reichsanstalt (Imperial Institute for Physics and Technology, 1887) and the Kaiser-Wilhelm-Gesellschaft (1911), both of which were closely connected with industry. The Technical University and the Prussian Academy of Sciences were founded, along with numerous museums and libraries, technical colleges, and military training institutes. Moreover, in the last quarter of the nineteenth century, the economic and social structures of the city were substantially changed by the establishment of mechanical engineering, electrical, and chemical firms, such as AEG, Siemens, and Schering.

Science, in this period, was closely linked to its urban context and figured prominently in the public domain. A movement arose to popularize science, starting with the founding of the Urania education center by Wilhelm Foerster and Wilhelm Meyer in 1888. Three *Volkshochschulen*—adult education institutes—were opened between 1878 and 1902, and a zoological garden and a botanical garden were created. This penetration of science into the public domain, including such developments as a sewer system and the introduction of standard time clocks, also represented a contribution to the dominance of bourgeois culture and power.

Driven by urban elites, Berlin's evolution into a modern metropolis was characterized by conflicts between the Kaiser and the administrators of new scientific and cultural institutions, such as the natural history museum. Although no international exposition was held in the city, the establishment of industry, the founding of scientific institutes, and the popularization of science paved the way for the introduction of a modern industrial lifestyle that integrated a vision of the traditional past with visions of the industrial present and of future national progress.

The growth of Edo (Tokyo) held some parallels with that of Paris in the seventeenth and eighteenth centuries. Edo was the seat of power for the feudal government under the Tokugawa family. With the Meiji Restoration of 1867–1868, the emperor of Japan was returned to a more central—albeit symbolic—role in leading the nation. Japan embarked on an ambitious program of rapid modernization that involved the large-scale introduction of Western science and technology. More than ever before,

Tokyo became the center of economic and cultural activity in Japan. In examining Tokyo's transformation in the years from 1868 to the period of reconstruction immediately after the Great Earthquake of 1923, this section outlines how Western organizational models were introduced during the Meiji era (1868–1912) and beyond.

Scientific experts have been important as shapers of urban space, not least because of concerns regarding public health. One of the most important figures in the transformation of Tokyo was the physician, colonial bureaucrat, and one-time mayor of Tokyo Gotō Shinpei (1857–1929). The chapter examines Gotō's ambitious vision for Tokyo, and the way in which he built up his expertise via contact with Charles A. Beard, a former director of the New York Bureau of Municipal Research, who was invited to Tokyo to help establish the Tokyo Institute for Municipal Research in 1922, just prior to the earthquake in 1923.

For Tokyo's elites, modernization was not only a matter of introducing Western institutions and values; it also involved presentation. Exhibiting the modern through expositions and museums in Tokyo helped to educate and enlighten the Japanese public and helped to shape the city as the capital of the nation. Beginning in the mid-1870s, expositions were held in major cities throughout Japan. Japan's participation in the 1873 world's fair in Vienna gave the Japanese useful experience, which they applied in mounting the First National Industrial Exhibition, held in 1877. The Ministry of Education Museum was launched in 1872, followed by the Museum of Education in 1877. In a way, museums were expositions made permanent. These cultural institutions, along with the cityscape itself, played important symbolic roles in the creation of Japan's modern identity, as well as its sense of its past, present, and future as a nation and an international power.

It should help readers if we explain here the meaning we give some key terms used in this book to describe the period we cover. They have taken on a life of their own, becoming something like scholarly buzzwords with vague associations. We have worked to re-moor them to the people, institutions, and events we examine here.

We consider "modernity" as a condition of existence whose major feature is acceptance of historical change as a given. As sociologist Anthony Giddens proposes, this condition exists at the institutional as well as the individual level: "It is a society—more technically, a complex of institutions—which unlike any preceding cultures lives in the future rather than the past."[6]

During the period covered by this book, we suggest that this sense of living in the future became rooted in people's lives through the rebuilding of cities, the mounting of expositions, and the forming of museums and the institutions that enable them. These activities were generated in a particularly urban context, in the public and private institutions, agencies, and organizations where elites were trained or engaged in setting standards, designing, building, and administering.

And just who were these elites? What were their objectives, and what means did these urban actors use to inaugurate and realize change? More broadly, how do we define the nature of the social relationships among these urban elites, which came to sustain a new international industrial culture?

We use the term "elites" to refer to the individuals who created and coordinated this culture. In a period of remarkable opportunity and profound social disarray, this new class of men had the power to demolish and rebuild cities, fund and organize vast expositions, and found and organize museums. Thus, they reframed the existence of vast numbers of people during the period of the second industrial revolution. Depending on their political and economic circumstances, these elites could be kings, emperors, elected officials, bureaucrats, industrialists, scientists and engineers, architects and planners, amateurs and professionals. Nevertheless, this galaxy of personages in Paris, London, Chicago, Berlin, and Tokyo shared a common desire to reestablish their societies in accordance with a set of modern premises on which they generally agreed. And they all focused attention on the urban setting as the space in which to begin this reframing.

Because even within particular urban contexts these elites were not necessarily linked by close friendships, political affiliations, or memberships in the same offices or businesses, we have chosen to use the term "nebula" to describe the historically novel character of their relationships.[7] From an international perspective, concurrences and convergences in their points of view derived in part from simple awareness of what was happening elsewhere, as well as the desire to emulate, compete, and solve similar problems: Napoleon III carried back to France his impressions of London's Hyde Park and the Great Exposition; the Chicago Commercial Club sought to best

Paris. Municipal and national governments sent official delegations and professional organizations traveled to investigate the sewers and metro systems of other world cities, and to visit their technical schools, expositions, and museums. Especially after 1880, some of these elites helped form new international professional associations that brought some of the second generation into a network of specialists working in urban planning, industrial education, scientific or technological research, and social hygiene institutions. The international expositions hosted congresses where these international umbrella organizations helped organize the efforts of many separate groups and individual elites.

In each city, this nebula had luminous patches of elites. Sometimes they congregated around a club, such as the Commercial Club in Chicago; in other cases they collaborated with one another across organizational boundaries for specific projects, as was the case with members of the Japanese commissions and those in charge of the international expositions. While some groups included representatives of the state or municipality, or otherwise could depend on relationships with those in power, in many instances the modernizing elites were seemingly isolated from the public sector, as were the impresarios who mounted the commercial expositions in London. Individuals acting on their own, or of different political persuasions, also came together solely to found new administrative arrangements, new institutions, and organizations to bring about progress in the urban realm.

However dispersed or clubby these elites, we consider science and technology as the means they used to thread their projects together and to resolve the historical difficulties they faced. They saw knowledge as a fundamental resource in rebuilding cities, running expositions, and founding museums, and ultimately in the creation of a new industrial order. Thus, by examining the role these leaders assigned to science and technology in the urban setting, it is possible to understand to a large degree what it means to say that this modern culture was based on science and technology.

Both terms were multivalent. Science in the era of Comtean positivism most commonly meant the systematic gathering and classification of evidence according to certain principles in order to put it to useful ends. For example, during this period, discoveries in thermodynamics and Darwin's theory of biological evolution painted nature as a dynamic system, always in the process of changing. Technology, understood by elites as applied science, was the means of actually using scientific knowledge to transform the material environment of the city into an entirely new ecology. This was also an era of dramatic technological breakthroughs in communications, transportation, construction, industrial production, mass consumption, and public health, which were integrated in a wholly invented urban fabric.

To see how theories of natural progress framed elites' projects, we will look at their plans for cities, expositions, and museums, as well as their interest in founding schools, research institutions, and facilities in the city to educate professionals to continue these pursuits. Moreover, in this era when sociology and anthropology emerged as academic disciplines, urban elites saw the value of putting that knowledge to work in museums and expositions to explain the past and shape the future.

Modernity also entailed a new awareness of time and space. Certainly, contemporaries thought of the decades after 1850 to 1870 as a historical watershed. In the past lay an agricultural society, primarily dependent on human, animal, and natural power, dispersed geographically and economically. What we today think of as "modern" individual technological inventions may have been abundant previously, but the railroads, the factories, and steamship lines had yet to be integrated into systems. The revolutions of 1848 in Europe and the civil war in Japan, the wars that created modern Germany, the Paris Commune, and the labor unrest in London and Chicago destroyed the old systems and put these societies on the verge of anarchy. By the 1870s the old societies were artifacts, the present incoherent. In contrast, science and technology measured out progressive time as the orderly path to a better future realized in the spaces of cities. Time was, in a sense, manufactured along with the geographic locations that were rebuilt. In the refreshed, modernized city, it was possible to believe that progress had materialized. The city was tailored to change, for it was capable of continuously integrating inventions, individuals, and institutions over extensive geographic areas into coordinated systems. Over time, this sense extended beyond the city, to the nation and internationally.

Why and how did this commonality come into being, and who was responsible for it? A great deal of important historical work has been done on this period in the form of biographies of major figures or studies of individual cities, expositions, and museums. Yet these studies have failed to note the unifying function of science and technology in defining and steering the development of these widely dispersed social spaces. We believe we are breaking new ground in two ways: first, by exploring how these geographically distant industrialized centers are similar, interconnected, and different; second, by employing a research model that introduces the comparative mode and the collaborative work of five scholars. While significant studies have identified a variety of economic and social factors leading to the changes we discuss, the role played by urban elites' new ideology of progress in establishing, shaping, and integrating the institutions of change in these centers has yet to be considered fully. The following chapters, each devoted to a different city, together take up this challenge.

NOTES

1. In 1871 the population was already over 3.26 million, rising to 4.39 million by 1931. If the suburbs of Greater London are included, the population rise was even sharper, doubling from 3.8 million in 1871 to over 8 million by 1931.

2. William Cronon, *Nature's Metropolis: Chicago and the Great West* (New York: Norton, 1991), 263–309.

3. John Glessner, *The Commercial Club of Chicago: Its Beginning and Something of Its Work* (Chicago: privately printed, 1910), 13–14, 46, 131–136.

4. Robert Wiebe, *The Search for Order 1877–1920* (New York: Hill and Wang, 1967), 129.

5. Donald Collier, "Chicago Comes of Age: The World's Columbian Exposition and the Birth of the Field Museum," *Bulletin of the Field Museum of Natural History* 40 (1960): 2–7, 4.

6. Anthony Giddens, in Giddens and Christopher Pierson, *Conversations with Anthony Giddens: Making Sense of Modernity* (Stanford: Stanford University Press: 1998), 94.

7. Christian Topalov, "Les 'réformateurs' et leurs réseaux: Enjeux d'un objet de recherche," Laboratoires du nouveau siècle, La nébuleuse réformatrice et ses réseaux en France, 1880–1914, 1999, n.p. (no longer online) http://www.ehess.fr/Editions/ouvrages/9914-Introduction.html#entete.

2 BRINGING THE FUTURE TO EARTH IN PARIS, 1851–1914

MIRIAM R. LEVIN

PART I   SETTING THE SCENE

Jules Verne's unpublished novel *Paris in the Twentieth Century* (1863) is the tale of an aspiring young poet adrift in a future metropolis monopolized by science and technology.[1] Pneumatically driven elevated trains, electrically lit streets, homes fitted with mechanical gadgets, libraries, research and educational institutions reflect and promote profitable investment in urban industrial progress. The city is run by an interlocking oligarchy of state administrators and capitalist bankers, who keep institutions on track by applying science and technology to urban development, record keeping, and accounting. A vast corporation called "The Academic Credit Union" oversees cultural innovation and its dissemination for instruction and amusement. At annual festivals held in its modern precincts on the Champ de Mars, the Union awards prizes to graduates for useful contributions to scientific and technical fields. A few are reserved for the arts—considered refreshing diversions for a hard-working population in thrall to industrial profit-making. In all, the future Paris is a drearily cheerful techno-system where "construction and instruction are one and the same for businessmen, education being merely a somewhat less solid form of edification."[2]

Like all good writers of science fiction, this popular inventor of the genre signaled the salient features of his time. Writing in the midst of Napoleon III's extensive rebuilding of Paris, Verne put his finger on two major developments in the city that continued to evolve under the early Third Republic: first, the assumption of a central role by elites in creating and administering new Paris-based institutions, organizations, and bureaucratic offices that structured modern industrial society; and, second, the commitment of these leaders to steering society along the path of progress via scientific and technological advances inaugurated in Paris. Especially important is the way they used a cluster of mutually reinforcing activities, in which these advances were embedded, to extend the geographic and imaginative reach of the new urban culture. For Verne's elites, these projects centered on infrastructure building, festivals, and scientific and technical education in the city. Similarly, the men of the Second Empire and Third Republic combined urban rebuilding (including the erection of laboratories and schools), universal expositions, and museums to restructure Parisians' reality. The appeal of these activities lay in their ability to simultaneously symbolize and materially advance an industrial capitalist system.

This chapter examines the activities of three generations of men who inaugurated, developed, and extended this modern culture of change in Paris over the latter half of the nineteenth century and the first fourteen years of the twentieth.[3] Readers are right to note that these generations span two regimes and two very different forms of government. There are good reasons, however, why the following pages consider them together. Overlaps in agendas, means, and personnel argue for establishing the continuities marking the advent of this culture in Paris—the city that Walter Benjamin baptized "the capital of the nineteenth century." Looking at distinctive differences between the forms of government, the policies regarding the Catholic Church, and the objectives of elite groups help account for changes in the character of urban modernity in Paris in the late nineteenth and early twentieth centuries.

Paris-based elites wished to master the forces of industrialization by constructing a set of science- and technology-based institutions, values, organizations, and spaces. The urban environment of the nation's capital became the site where economic uncertainties and social strife would give way to a future seemingly insured against such difficulties. In Paris a controlled sense of the present materialized in rational aesthetics of open vistas, logically proportioned buildings, green parks, and squares, while visible and hidden improvements to the city's infrastructure orchestrated a host of new social and economic relationships. The plans city leaders devised for the universal expositions were designed to symbolize and inaugurate the creation of ideal industrial societies, and broaden the international cooperation thought necessary for their realization. Materially, the fairs' construction and servicing offered government officials periodic spurs to continue developing Paris's infrastructure and its neighborhoods, as well as to stimulate its economy. While the fairs seem to have been focused on the future, they also fed the population's growing awareness of the past as governments moved exhibit collections to existing museums and founded new museums in fair buildings, increasing the presence in Paris of these institutions housing the past. Additionally, under Napoleon III and especially under the republicans, museums, laboratories, schools, and the university extended their role as sites in the city for scientific and technological research and education.

We can understand these continuities—and shifts of focus—in cultural institutions by looking at the networks formed by the men responsible for them. Despite the change of regimes from empire to the democratic Third Republic, there was a marked continuity of individuals and sources of personnel engaged in planning and realizing urban rebuilding, universal expositions, and museums. While not always linked by close friendships, party affiliations, or business connections, these men were part of the nebula of reformers and rebuilders discussed in the introduction. Coming from government, business, manufacturing, and finance, they belonged to a variety of private, public, departmental, municipal, national, and international organizations and institutions. Together, these men and organizations constituted a loose system of progressive forces that began to form as France, after 1851, joined her Atlantic neighbors in a period of serious industrialization.[4]

Within this framework, significant shifts took place in the character of modernity. During the 1850s and 1860s, Emperor Napoleon III's goal was to turn Paris into a hub of industrial capitalism, where investments in urban development would stimulate economic growth and general prosperity.[5] Ceding control of the university and schools to the Catholic Church in exchange for its political backing of his regime, the emperor still had a wide set of possibilities open to him as he centered power around himself. To assist him, he appointed a cadre of like-minded administrators, engineers, and bankers to various commissions and government posts. Prominent among them were his extremely able prefect of the Seine, Georges-Eugène Haussmann, and a number of graduates of the École Polytechnique (EPT), some of whom were strongly influenced by Saint-Simon's social philosophy.[6] After assuming his position in 1853, Haussmann marshaled the talent and resources to realize this vast, coordinated project. He looked to a state-employed engineering and architectural elite to design and administer this transformation through private contractors and free labor.[7] Directly responsible to him were engineers F. E. Belgrand and Adolphe Alphand, in charge, respectively, of the Service des Eaux-Égouts (Water and Sewers) and Service des Promenades et Plantations (Sidewalks and Plantings; later la Voie Publique et les Promenades). Alongside them were architects from the imperial École des Beaux-Arts, including Gabriel Davioud; Ernest Labrouste, architect of the iron- and glass-domed Bibliothèque Nationale, widely regarded as the most innovative work of this generation; and Charles Garnier, architect of the Paris Opéra, one of the keystone structures in Haussmann's Parisian vista. Together they established standardized guidelines and systematic working relationships on which the construction of the new urban infrastructure was predicated.[8]

They also used universal expositions in the capital to advance the emperor's building projects, as well as to celebrate industry and material progress itself. Gestating within the exposition of 1867 were seeds for future social innovation planted by two of the emperor's *équipe*: the sociologist Frédéric Le Play and minister of education Victor Duruy. These men concerned themselves with the social and economic end of things—Duruy undertook reform of education and research while the *polytechnicien* Le Play dealt with economic policy and expositions, aided by his friend economist Michel Chevalier, member of the Council of State and also a *polytechnicien*, and by his protégé Jean Jacques Émile Cheysson (1836–1910). Under the Third Republic their reform projects on display at the exposition would come to fruition respectively in the Musée Social, founded by Cheysson, and in the expansion of science education.[9]

Although Napoleon III and Haussmann tended to look forward rather than back, they found it necessary to respond to the public's dismay at seeing so many old buildings and streets—and the past itself—vanish before its eyes. Hence, by 1860 the emperor began to actively support the Service des Monuments Historiques, putting at its head a favorite of his, Eugène-Emmanuel Viollet-le-Duc, preservationist and architect of the new Palais de Justice, one of the centerpieces of Haussmann's Paris.[10] For his part,

Haussmann founded the Musée Carnavalet, dedicated to the history of the city of Paris.[11]

By contrast, the aggressively laicized Third Republic amended the new culture to accommodate a parliamentary system with a strong democratic and moral reform agenda for industrial capitalism in France.[12] Within the dispersed power centers of the new government's bureaucracy, elected officials and appointed personnel (some of them carryovers from the Second Empire) slowly continued construction projects in Haussmann's plan for Paris (figure 2.1).

Added to these efforts was the long-awaited Métro, built under the auspices of a newly independent, left-leaning Paris Municipal Council. Having ousted the Catholic Church from control over the education system, the republicans were keen to add and refurbish schools, university teaching and research facilities, and museums as means of creating self-directed, improvement-minded citizens. Expositions and museums also absorbed their attention as places in the city where the past, present, and future became connected into a narrative of continuous republican progress.[13] Liberal and democratic Enlightenment ideas, particularly those of Condorcet regarding the social and moral utility of scientific knowledge in a free society, fed their commitment to these institutions and activities as correctives to oligarchic capitalist ends.[14]

This culture of change had its last reworking in the decade prior to World War I, as the automobile, the airplane, and large-scale industrial production came into play. Haussmann's plan had reached obsolescence, and the political will for further urban improvements faltered. In their place, a generation of professionals using a wide range of statistics, measurements, and methodologies focused their expertise on organizing museums and drawing up a new set of plans for a redesigned Paris. Working in public and semiprivate institutions in the city, including the Musée d'Ethnographie du Trocadéro and the Musée Social, these men turned their attention to organizing the past and future of the city.

Napoleon III and his *équipe* were men of what I call "salutary urban communications," intent on putting human beings and a modern built environment into a smoothly functional, healthful, and aesthetically attractive interactive network. "Communications" refers not simply to the material networks of telegraph, boulevards, transport, and gas and electric illumination that improved the circulation of people, goods, and capital,

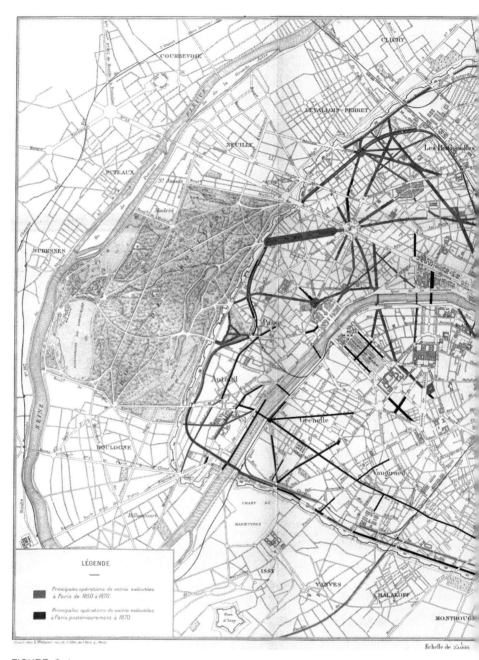

FIGURE 2.1

Paris. Map of Paris showing streets constructed between 1853 and 1914. From Commission d'Extension de Paris, *Considérations techniques préliminaires* (Paris, 1913).

Photo courtesy of African Studies Library, University of Cape Town Libraries, Cape Town, South Africa.

but to a means of institutionally creating and rationally managing dynamic economic and social relationships within the city, connected nationally and internationally through an urban environment built out of the most advanced technologies available. They legislated a rational basis for aesthetics linked to functionality and the integration of new materials and methods of construction.[15] Through their centrally directed projects, they coordinated government, private banks, construction firms, realtors, and the vast working populations of Paris in a set of interdependent relationships, while rationally reconfiguring the social spaces within which residents and visitors consumed the fruits of this construction. In turn, building and living within these environments engendered dynamic changes in physical, material, and psychical existence. International expositions and museums arose from the commitment to rebuilding Paris and served as stimuli for further improving and extending this way of life into new spaces and to more people.

The impact of the emperor's efforts was profound; his vision, however, could not be fully realized. The economics didn't work, and so construction slowed, stopped, or occurred erratically.[16] The growing and increasingly segregated working classes did not share in the benefits of the transformed city. The hold of the church and of loyal Catholics over instruction blocked the formation of a new generation of scientists, engineers, and technical personnel trained in research and modern applications to replace the old cadre. And finally, the static nature of the goals to which the emperor directed change, combined with the failure of his efforts to liberalize the political system, left the future of the culture of change at an impasse by 1870. The war with Germany, the siege of Paris, and then the formation of the Commune necessitated a recasting of this past effort, though not a complete break. From the late 1870s until the outbreak of war in 1914 under the Third Republic, the loose associations between the development of the city, expositions, and museums would be tightened, rationalized, and reformulated into an open-ended narrative of progress, along with conceptions of the present, past, and future based on science and technology.

As sketched in this essay, by examining the development of this culture of change under the early Third Republic, we may see that democratic configurations of scientific knowledge and technological design left their own stamp on modernity. Moreover, by giving a larger place to science and technology as factors in the systematic relationships that these men created among urban rebuilding, museums, and expositions, we may better understand the new conception of constructed time and of the possibility for its control that became characteristic of the modern condition. Finally, by considering the virtual standstill in Parisian construction after 1900 in light of that period's professionalized planning based on the automobile, airplane, electricity, and large-scale industry, we may understand how the future orientation of modernity itself was reified into a set of specialized practices for dealing with contingencies arising out of the creative dynamic of capitalism itself.

PART II   THE THIRD REPUBLIC: A REPUBLICAN RIFF ON THE CULTURE OF CHANGE

It was Paris that reigned with sovereign sway over the modern era, and had for the time become the great centre of the nations as they were carried on from civilization to civilization, in a sunward course from east to west. Paris was the world's brain. Its past so full of grandeur had prepared it for the part of initiator, civilizer and liberator. Only yesterday it had cast the cry of Liberty among the nations, and to-morrow it would bring them the religion of Science, the new faith awaited by the democracies.

—Émile Zola, *Paris* (1898)

Defeat at the hands of the Germans, two long sieges of the city, and the establishment of the Paris Commune left Napoleon III's republican successors facing an anxious, insecure, and socially divided population. Leaders of the new republic also inherited an even more difficult legacy from the defunct empire—one that challenged their commitment to creating a progressive liberal democracy through science and technology. The international network of manufacture and trade now constrained the population within a centralized system of circulation that left little hope of controlling its own destiny. Haussmann's Paris itself was source, product, and symbol of these imperial policies for managing change. For the republicans, rooted in the Enlightenment and the democratic ideals of 1789, progress would not be so simple to attain as it had once seemed. Paris, and through her the country, would have to accommodate republican visions with the realities of a modern industrial society.[17]

Jules Ferry, minister of education and fine arts and author of the 1880s laws that made French primary education free, universal, and secular, argued that the government would have to work within the now technologically integrated capitalist system: "It is absolutely impossible for our industry to avoid the prodigious, and I know, often painful and depressing effects of those means of communication which have turned the entire world into a single marketplace."[18] If the republicans reconciled themselves to existence within an international web of railroad, steamship, and telegraph companies, their spirits were buoyed by their belief that the quest for riches would be second to that of individual and community betterment. Properly used, science and technology would enable French citizens to capture the energies of industrial capitalism for their own and the republic's benefit. This, however, would be an evolutionary, not a revolutionary process. Among leaders of the 1880s, Jules Simon—always ready with republican catchphrases—made this clear in rather idealized language: "We should finish in freedom what was begun by despotism."[19]

Unsurprisingly, then, there were certain continuities with the Second Empire. Prominent among these was the presence of certain key individuals who had served the emperor. Moreover, republican elites also relied on the established dynamic of urban rebuilding, elaborate expositions, and museums to aid them in their efforts. And they continued to realize Haussmann's rebuilding plan, which they saw as essential to urban expansion, public health, and circulation. By the early 1900s, they had brought the Haussmannian conception of the modern city to its logical conclusion.

However, there were significant innovations. First among them was transforming the city from an imperial into a republican capital, identifying science and technology with liberty and democracy. Second, with finances lacking and multiple political responsibilities, officials came to rely even more heavily than they had during the empire on expositions and museums to advance urban development. And third, at the turn of the century, the advent of electricity, the automobile, and the airplane, coupled with persistent urban growth, led a new generation of administrators to reinvent the culture of change to accommodate a sleeker, faster, more efficient vision of urban modernity. Not to be overlooked is the reorganization of government, the increased bureaucratization and number of civil servants that oversaw these projects in the city.[20] By breaking the Catholic Church's control over education, republican elites in the national government and administration made Paris the center of a public school system for inculcating their secular ideology of material and moral progress, constructing laboratories and institutions in the city where new cadres of scientists, engineers, and technicians would continue to lead the republic forward.[21]

After 1890, the city of Paris was granted an elected municipal council for the first time since 1849. The growing prominence of socialists in the council affected planning, funding, and construction of the city's first major public transport system: the Métropolitain.[22] Also important in the 1890s and 1900s was the promotion of urbanism by nongovernmental associations and organizations, legalized under the new freedom-of-association laws. As Calabi has written of the town planning movement that emerged at the end of this period in which Musée Social members participated, "the links were neither institutional, nor programmatic but practical."[23] Men in these associations established relationships among various government bureaus, expositions, and museums, eventually developing a new vision of Paris that reinvented the culture of change itself.

In place of Napoleon III's vision of the city as the center of an empire of the wealthy, whose networks of railroads, steamships, and Saint-Simonian technocracy drew all classes into its orbit, the republicans enlisted science and technology to make that system support individual initiative and a quest for social improvement on which democracy could pin its hopes.[24] In the republicans' view, a scientific and technological Paris was not only the spearhead of republican progress, but of the world at large. The passage from Émile Zola's 1898 novel *Paris*, quoted at the beginning of this section, captures their dynamic urban ideal perfectly.

In sum, the Third Republic would not throw out the achievements of the Second Empire in Paris, but would alter the character of modernity in the city to make it serve the middle class. Its leaders would introduce their own political agenda for science and technology, one that would liberalize and democratize the culture of change. Renewed urban development in conjunction with international expositions fed a growing belief that progress was an evolutionary process toward republicanism—the apex being the 1889 exposition marking the Great Revolution's centennial. Museums, where evolutionary theory would come to frame research and exhibition design, turned the past into a prologue for the republican present and future. By the turn of the century this approach to managing change had matured. The new generation who managed it, however, began to feel that new goals and new technologies were needed to accommodate the city and its inhabitants to the elaborate industrial system that Jules Ferry wished to tame.

THE REPUBLICAN NEBULA

Under the Third Republic, the nebula of individuals contributing to the Parisian culture of change was much more diverse, its leadership more decentralized, and its power more dispersed than previously. Its leaders and main support came from the *moyenne bourgeoisie*, not from the aristocracy, the *grande bourgeoisie*, bankers, or Catholic hierarchy of the Second Empire. Raising funds both via taxes and government/business collaborations, they avoided the arrangements with private banks that had been Haussmann's undoing.[25] These elites were lawyers and other professionals, manufacturers, intellectuals, teachers, civil servants, and reformers committed to placing the republic on a scientific foundation. Some were Protestants, and some of the key members had been politically active in the opposition during the last years of the empire. In the late 1890s a second generation of professional elites emerged, calling themselves urbanists. The urbanists argued that geographic expansion, a new social geography, increasing mobility, and new industries of electricity and transport (bicycles, automobiles, and airplanes) required yet another rebuilding of Paris to accommodate the fully industrial era that had arrived.

The elites of the republic worked within a number of reorganized and newly founded ministries and government offices. They were creating the world of rational bureaucracy that their contemporary, the German sociologist Max Weber, would so astutely analyze. From the outside it might appear that those involved labored in the grip of oppressive scientism, but viewed from the perspective of its creators, as Weber pointed out, participating in the making of this new coherent order was enormously energizing and liberating.[26] The Prefecture of the Seine was reorganized, and the Paris Municipal Council became an elected body independent of the prefect. The now laicized Ministry of Education was combined with that of Fine Arts, with control over schools and museums. Among its charges was wedding the arts, architecture, and urban design to scientific, technological, and secular moral education.[27] As already mentioned, the administration of the promenades and streets was divided into two specialized departments. In a number of cases, these elites used their posts within expositions, museums, and research institutes to advance their specialized interests.

Almost all had degrees from universities or the state professional schools, in arts, architecture, literature, history, law, and engineering. Their biographies also show that they were active in a number of nonprofit and emerging professional associations. But, as one would expect in a multiparty democracy taking shape over a thirty-year period of intense innovation and industrialization, personnel changed over time, bringing with them different institutional agendas and positions on how to marshal science and technology in the development of the city.

The elites of this time had a view of progress that was strongly grounded in history, particularly that of the Enlightenment and French Revolution.[28] In their account, scientific knowledge itself evolved in a positive direction and along with it individual consciousness and human society. If the government steered the society along its proper course, it was scientists who provided the knowledge to be used to make the nation strong and sure.[29] They had in common a belief in liberty, equality, and fraternity; although they ranged from radical to left to conservative, these republicans all held to the Enlightenment's conception of the rational individual as the building block of society. Like their eighteenth-century intellectual predecessors Condorcet, Condillac, and Diderot, they understood that through a scientifically based education, in and outside the classroom, rational nature would produce self-directed citizens who would orchestrate their actions in common.[30] These actions were both self- and society-building, and they were associated in a deeply reciprocal way with constructive labor and the constructed environment.

In an age long before interactive computer games and electronic learning, these republicans considered the aesthetic character of the environment a significant factor in this educational experience. Design shaped the people who created and lived within these scientific environments, who then built and shaped what were essentially control technologies to serve their own ends.[31] Hence, the republicans added moral authority to the economic and political significance Napoleon III had attributed to urban settings, expositions, and museums. A designed environment incorporating scientific knowledge and processes, as well as new technologies, constituted reciprocal ways of creating social cohesion and individual initiative. In short: democratic progress.

Overlapping clusters of elites within the republican nebula included a number of politicians, some of whom had honed their skills during early careers in the last years of the Second Empire. In the 1880s ministers like Jules Ferry (credited with the new lay, free, and universal primary education system), Jules Simon (an advocate of improved technical education and the founding of industrial schools), and Édouard Lockroy (a radical republican who married into Victor Hugo's family) oversaw initial planning for the 1889 centennial exposition and the selection of Gustave Eiffel's tower as its defining monument.[32]

Two politicians, standing at either end of this period, were closely linked to private-sector efforts at urban improvement. The elder, Léon Say (1826–1896), was a liberal economist who had established his reputation during the Second Empire through a series of published attacks on the financial administration of Baron Haussmann. In the Third Republic Say served in a variety of elected and appointed posts and used his talents to popularize the science of economics. After serving as prefect of the Seine in 1871, elected member of the Senate, and finance minister under several administrations, by the early 1880s Say recognized the need for government to support public works. Thus, he came to head the commission that organized the Social Economy section of the 1889 exposition, working with Cheysson to found the Musée Social, which grew out of that effort. In the following decades, Jules Siegfried (1837–1922), Say's younger colleague, worked with him in this effort.[33]

Outside the ranks of politicians were men appointed to administrative posts. Trained in the law, Eugène Poubelle (1831–1907) served, like Haussmann before him, as prefect of the Seine, from 1883 to 1896, and left a significant mark—if not so great as his predecessor's—on Paris. Following an outbreak of cholera in Paris in 1892, Poubelle successfully pushed the government to accept the policy of *tout à l'égout* (all waste in the sewer). A law was passed in 1894 requiring all Parisian property owners to hook their buildings to the sewer, and to pay a tax for garbage collection and sewage and water service. Since then Poubelle's name has been synonymous with the garbage cans his legislative efforts required all Parisian property owners to maintain outside their doors. Less well known are his efforts to increase the scientific workforce by opening medical internships in Paris hospitals to women.[34]

Polytechniciens from the École des Ponts et Chaussées continued to play a prominent role in extending and further embedding the culture of change. Although Saint-Simonianism among these men had attenuated with time, traces remained in a strong sense of esprit de corps associated with careers dedicated to progress in the service of the public. In fact, engineers in state service now had the same status as bourgeois industrialists, bankers, and magistrates. César Daly, editor of the *Revue générale de l'architecture et des travaux publics*, characterized these designers and administrators of the urban infrastructure as the great creators of "la richesse publique" (public wealth) on which modern civilization was built.[35] These men were attached to the Ministry of the Interior, which maintained the simplified and unified organization of the service established by Haussmann. Through extensive travel to observe urban developments abroad, they brought an informed and international perspective to their work.

Among this cadre, a major force for continuity in their approach to cultural change was Adolphe Alphand, who served under the prefect of the Seine. After 1871, he became the powerful head of the newly renamed Service de la Voirie Municipale de Paris. In that capacity, he drew together the statistical services of the state and a team of architects, engineers, and horticulturalists to survey and refurbish the damaged city and to continue to pursue Haussmann's plans for it, as well as to handle the construction of the expositions of 1878 and 1889. His calm assurance, bourgeois professionalism, and hands-on approach to administration are evident in a portrait in which he wears a top hat and dusty—but proper—dark suit at the construction site of the 1889 exposition (figure 2.2).[36]

Alfred Picard (1844–1913), a graduate of the Ponts et Chaussées who had managed the development of eastern railways and canals, was equally impressive. In the 1880s he headed the Ministry of Public Works section in the Council of State. At the 1889 and 1900 Paris expositions, he put these talents to work, first as editor of the final report and then as chief commissioner in charge of the planning and realization of the 1900 exposition, including its extension onto the Right Bank, allowing for further development of the lower part of the Champs-Élysées. Deeply familiar with the development of the national rail network, Picard appreciated the influence public works could exert on popular acceptance of state-supported science and technology.[37]

Fulgence Bienvenüe (1852–1936), another graduate of the EPT and the École des Ponts et Chaussées, working for the city council's Paris Section Municipale de la Voirie Publique, brought the integrated, efficient systems approach of the *polytechniciens* to bear on the transportation, hygiene, and

FIGURE 2.2
Paris. Portrait of Adolphe Alphand on the construction site of the 1889 Paris Exposition Universelle, by Alfred Roll, 1888. Collection Petit Palais, Paris.

illumination systems of the city. His most important accomplishment was the design and construction of the massive Métro system begun in 1898. Following the wishes of the Paris Conseil Municipale's more left-wing members, he devised a system that represented a break with the railroad interests, which had hoped for over forty years to extend their networks into the city itself.[38]

Within this nebula, star in his own solar system, was Gustave Eiffel (1832–1923), private entrepreneur, successful graduate of the École des Arts et Manufactures, and probably the most famous engineer of the period, due to his famous tower. He was also the best known of the private contractors who helped realize government and real estate developers' agendas for Paris. In effect, he was the ideal republican citizen, an upright individual who used his knowledge and talents to benefit himself, the economy, and the nation. Yet, as a player in the creation of an organized liberal society, Eiffel himself was neither a republican moralist nor a social reformer. Rather, he was a practitioner with a symbolic aura—an enactor of technological and scientific progress, putting the laws of nature to work for human ends. To his mind, his iron bridges, domes, buildings, and the Eiffel Tower itself brought the order of natural law into the realm of everyday life to make it more efficient and more beautiful. He had adapted his designs and manufacturing methods from those he had used in constructing railroad bridges, bringing them into the heart of the city and making expensive masonry construction outmoded. And although it may seem that by the time of the 1900 exposition iron struts were getting to be old hat, Eiffel managed to use his giant structure to launch radio transmissions and experiments in aeronautics, helping tame technologies of the future. Thus, in making the transition from the Second Empire's politics and technologies, he was not only a liaison between public and private worlds, government and the wider public, but a consistent voice expressing how to manage industrial change through science and technology.[39]

Professors, new players who came to stand at the educational nexus of this culture of change, infused new meaning into the role of science and technology in the city, the museum, and the exposition. Posts in the reformed educational system provided them professional status and places from which they could create further networks. This was particularly true of a cadre of academics engaged in urban issues, museums, and expositions who took posts in the newly reformed, secularized, Paris-based system of higher education. They used these venues to develop both contemporary projects and systematic approaches to their subject matter based on continuities they hoped to establish between past, present, and future. Among their ranks was the aesthetically conservative architect Charles Garnier, professor at the École des Beaux-Arts. The 1889 exposition gave Garnier the opportunity to display his archaeological erudition; he also used the occasion to place his preference for stone over iron construction in a historical context. There, in an extensive exhibit on the history of human habitation, he promoted modern housing as a measure of civilization's progress and a source of laws for contemporary domestic design.[40]

Much more radical in spirit were a cluster of men who held professorial posts at the Muséum National d'Histoire Naturelle. Besides contributing to the development of the human and natural sciences, they also provided enormous support for evolutionary theory as an explanation for change compatible with republican ideology. Zoologist Alphonse Milne-Edwards inaugurated the triumphant return of the naturalists in 1891 when the minister of public education named him director of the museum. Milne-Edwards's appointment ended an administrative policy in place since 1863 (the date chemist Eugène Chevreul began a long tenure at the museum, followed by chemist Edmond Frémy) that had emphasized the empirical research more acceptable to the Catholic Church.[41]

Under Milne-Edwards's aegis, Ernest Hamy (1842–1908), holder of the chair in anthropology at the Muséum, helped establish an evolutionary comparative approach to the scientific study of human activity, and particularly of that activity central to republicanism: work. Setting himself as an example on this count, through arduous diplomacy and intellectual effort Hamy was able to acquire anthropological and ethnographic collections exhibited at the 1878 exposition, and then to found the Musée d'Ethnographie du Trocadéro (a new division of the Muséum housed in the Trocadéro Palace). As that museum's first director, he created a center for ethnographic study and public education. Using his connections at these two museums in the capital, Hamy emphasized the progressive evolution and variety of human cultural production in an organized setting.[42]

More directly engaged in integrating workers and their families into the equation of urban reform were men affiliated with the Musée Social, including two professors.[43] This private organization, funded by the wealthy owner of the Baccarat crystal works, the Comte de Chambrun, was founded in 1894 as a center for research, archives, exhibits, and policy making, aimed at extending the benefits of the modern city to the working classes. Its members drew expositions and the development of the Parisian infrastructure into their sphere. The involvement of politicians Léon Say and Jules Siegfried in this important private institution has already been mentioned. Outside the halls of the Senate and Chamber of Deputies, the engineer Émile Cheysson (already mentioned as a disciple of Le Play), the bibliophile Marcel Poëte (1866–1950), and the architect Eugène Hénard (1849–1923) represent a range of approaches to the urban-centered culture of change focused respectively on the present, past, and future of Paris.[44]

Their efforts centered on what, after 1900, became the new specialized field of urbanism, given impetus by the new commission on the expansion of Paris and the international phenomenon of urban growth and the heightened pace of industrialization. Primary among these men was Cheysson, who had assisted Le Play in organizing the exposition of 1867. In the 1880s, he became Professeur d'Économie Industrielle at the École des Mines, and he led the organization of the social economy section of the 1889 exposition. Cheysson worked within an extensive network of official and private professional societies. A prolific writer, he produced wide-ranging works on contemporary economics and problems of working class housing in which he made extensive use of statistics. These included the *Album de statistique graphique*, an annual series he directed for more than twenty years, using all known graphic forms (including pie charts, line graphs, bar charts, and flow maps) to depict data relevant to planning (railways, canals, ports, tramways, etc.). He collected and published data to inform politicians and the public about contemporary urban issues and to draft legislative proposals including the *habitations à bon marché* (HBM) or low-cost housing laws sponsored by Siegfried, his associate at the Musée Social.[45]

If Cheysson focused his energies on pursuing the culture of change in present circumstances, a younger fellow activist in the Musée Social, Marcel Poëte, was equally dedicated to making Paris's past a scientific basis for a city more sociable and accommodating of industrialization. As systematic as Cheysson, Poëte's work was obviously very different in content, focus, and method from that of the *polytechniciens*. He conceived of the city as a living entity, continually recreating itself as it adapts to changes in the larger economic environment, while retaining traces of the past in its contemporary form. Thus, in his history *La promenade de Paris au XVIIe siècle*, he not only traces the origins of the contemporary Champs-Élysées and street configuration at the Étoile to the royal agendas of Louis XIV, but warns that the city will have to accommodate these spaces and uses to the contingencies of industrial society.[46]

Out of this evolutionary approach to the city, Poëte encouraged the study of the past as a way to find solutions for the future. To this end, he invented a set of categories for organizing archival documents, a bibliographic system akin to that used in the natural sciences. Poëte awkwardly bridged two emerging disciplines he was helping to define: urban history and urban planning. A recent scholar, Donatella Calabi, has aptly captured this dichotomy, calling the reformer both a "pioneer of Urbanism" and a defender of "the history of cities."[47]

Poëte was a graduate of another *grande école*, the École des Chartes (class of 1890), with an education rooted in archival research on institutions. In 1903 he was nominated conservator of the Bibliothèque des Travaux Historiques de la Ville de Paris, which had been established in the aftermath of the destruction of the Hôtel de Ville during the Commune. In this new post, he became absorbed in questions of educational experiment, offering the first course on the history of Paris to generate interest in this new subject among a large and varied public. As Calabi notes, "urban history became an object of civic and professional obligation, more important than mere scientific research."[48] His objectives went beyond scientific analysis, for he saw a new era of heavy industry on the horizon to which Paris would have to accommodate itself. Ever striving for practical results, Poëte used his membership in the Musée Social and a number of groups, including the Association de Vieux Paris and the Association of French Urbanists (founded in 1911), to help him focus and effect his goals.[49]

Beaux-Arts-trained architect Eugène Hénard (1849–1923) was equally aware that an integrated industrial society stood at Paris's doorstep. Hénard is today better known than Poëte, because of his essay "The Cities of the Future," presented at the landmark international Town Planning Conference in London in 1911.[50] In Hénard, we have an architect working within the state bureaucracy to solve problems in a manner that incorporated the kind of systems approach we have seen used by the *polytechniciens*. It stands in marked contrast to the approach of the older and more traditional architect Charles Garnier, which emphasized monumental statements over the functional integration of buildings into the urban fabric, for example.

Moreover, Hénard was sensitive to the solutions that technological innovations, particularly in transportation, could provide to growing demand for efficient circulation. The plans he drew up for various projects show him assuming responsibility for reconfiguring or even reinventing the city to incorporate such innovations.[51]

Hénard developed his approach to comprehensive urban planning and traffic circulation in part through his work on the expositions of 1889 and 1900. In 1889 he supervised construction of the massive Galerie des Machines, and in 1900 he designed the Palais de l'Électricité, and assisted Picard with the plan for the Grand Palais and network of roads connecting the Champs-Élysées with the Left Bank. He also responded to government calls for proposals to plan the expansion of the city and the development of its peripheral spaces in the 1900s, and to the Musée Social's program to incorporate low-cost housing into the industrial city. Between 1900 and 1914, Hénard produced a series of projects and studies that faced the challenge of continuing industrialization by wedding new technologies with the conception of the modern city as a center of national administration and international civilization. He brought his imagination to bear on inventions such as the traffic circle, constructed to his designs and based on contemporary studies and projections for increased urban circulation. Most importantly, he integrated the nascent technologies—the automobile, the Métro, the telephone, electricity, and air transport—in plans for a Paris that might be.[52]

Between 1880 and 1914 the men discussed here remained committed to developing the city of Paris, and to mounting expositions and establishing museums as sites of national change based on science and technology. If we now turn to the material examples and fruits of their efforts, we can track the shifts in the culture of change as it moved from a focus on construction to planning for a future based on the evolution of science and technology.

THE CITY OF PARIS

The Belle Époque was a memorable moment for Paris, as it seemed nineteenth-century modernity was about to realize its promise. Much of the building in Paris during the early Third Republic further embedded industrial capitalist interests in a rationalized system allied with the central government. Yet the republicans managed to integrate new directions for science and technology into urban construction during this period, slowly turning the city into a more middle-class, democratized environment.[53]

Haussmann's legacy continued to be felt, as rebuilding in the city followed his plans for streets, sewers, and parks and retained his guidelines for street width and building height. These calculated guidelines steered construction of the boulevards Saint-Germain and Henri IV and the radial roads around the Place de la République (figure 2.3). They also set the constraints for Charles Garnier's Opéra and the complex of streets surrounding it, which required removing all the medieval houses and lanes along with the entire hillock on which they stood.

Alphonse Alphand continued to supervise and design Parisian streets, parks, and urban furniture in the 1880s, finding opportunities to use the new paving materials that had been developed with increased traffic in mind. Before the 1890s, streets and roads paved with a mixture of stone and macadam covered 1.4 million square meters of the city. After 1887, stone paving was expanded, covering 73 percent of the city's streets, or 6.3 million square meters, by 1896. Engineers made the city into a materials-testing laboratory. They also benefited from exchanges with British and American engineers who were trying new road surfaces, including, by the 1890s, those suitable for automobile traffic.[54] The state established standards for a number of available pavement materials and took possession of a site at Senlis in 1878 to provide Paris with stone that met these standards.[55] This increasingly organized process meant that by the end of the century, all classes of Parisians moved within an extended system of streets that presented a generally uniform and organized, yet aesthetically distinctive vista.

Three times as many buildings were erected in the first decade of the Third Republic as had been in the last decade of the Second Empire.[56] These continued to have stone facades and shells, but iron began to be used more frequently in the supports. Such structures extended the proportional system of Haussmann outward into newly developed high-end housing in the 7th arrondissement around the Champ de Mars and into middle-class areas in the 17th, 18th, and 19th arrondissements in the city's outer ring. Everywhere the public went in these districts, they saw Alphand's hand in the regular placement of cast-iron lampposts and kiosks.

Within this mathematical web, Alphand also oversaw construction of thirty-seven gardens and squares, compared with Napoleon III's dozen or

FIGURE 2.3
Paris. *Demolition Opening the Way for Construction of the Boulevard Saint-Germain, Paris*, etching by Maxime Lalanne, c. 1870s. Library of Congress Prints and Photographs Division, Washington, D.C.

so.[57] What might be seen as spaces escaping the rigor imposed on the streets were in fact carefully calculated artificial counterpoints. From the layout of paths and planting of bushes and trees to the artificial lakes, grottos, and faux-wooden fences, the parks were built using standardized procedures and modern materials.[58]

Haussmann's identification of urban renewal with public health was elaborated upon by Poubelle during his tenure as prefect of the Seine. As mentioned above, after the cholera outbreak of 1892, the city modernized its sewers, leading to further development of the system as it reached into individual households. Although state-employed scientists and engineers published an increasing number of studies, the question remained of what to do with the output of the sewer system. While Baron Haussmann watched disapprovingly from the sidelines, heated discussions in the Chamber of Deputies, lobbying by Belgrand's successor Georges Bechmann, and a host of published studies by advocates on both sides led to a decision to use filtered sewage to fertilize suburban farms, rather than return polluted water to the river. This policy for handling waste demonstrated the use of science and technology to find a mutually beneficial accommodation between the needs of an expanding urban population and its suburban agricultural support system.[59] What more down-to-earth proof of this modern accomplishment could there be than deputy Martin Nadaud's report on his visit to the sewage-farmed fields of Grennevilliers? "I drank sewer water," he told those gathered in the Chambre.[60]

As André Guillerme has pointed out, under the aegis of the prefects and the municipal councils, engineers in government bureaus and private businesses contracting with the city increasingly surrounded Parisians of the central city with wires, pipes, and other conduits carrying measured flows of power for public and private uses.[61] Gas lines, serving since the 1860s to illuminate streets and larger commercial buildings such as department stores, were hooked into middle-class Parisian apartments where coal gas began to be used for heating, cooking, and interior lighting. Compressed air lines coexisted "with those for water and sanitation and with the new 'online' fluids: electricity and the telephone."[62] Electric power was slow to develop in the city, although Edison himself was something of a public hero. While the 1881 Exposition Internationale d'Électricité and the universal expositions of 1889 and 1900 helped introduce and create a demand for electricity, public electric services did not start until 1889, after they had reached the provincial cities of Saint-Étienne and Tours. The same lag applied to the telephone, exhibited first in Paris at the 1889 exposition.

Political concern also centered on turning urban construction and technologies to more inclusive democratic ends. As a result, expensive private horse-drawn omnibuses were transferred into public hands, and the city established new electric tramlines, a steam-driven rail line that circled the city, and the Métro. The Paris Conseil Municipal successfully insisted that it take charge of Métro planning and construction. Together these various forms of transportation provided a network—however expensive and imperfectly interconnected—of modern public transport that both kept the railroad interests out of the center of the city and made it easier, if not cheap, for Parisians to get back and forth between the center and the periphery of the city.

Within this network, national and municipal government and private initiatives constructed spaces supporting free development of science and technology. These interests encouraged institutions for research and education in science and technology of a republicanized generation of scientists, engineers, technicians, and skilled laborers. In part they pursued Victor Duruy's policies, now unhampered by political interference from the Catholic Church. From the 1880s through 1900s, the government constructed new laboratories and classrooms and refurbished existing ones in institutes and *lycées*, at the University of Paris (École de Médecine and École des Hautes Études), and at the Conservatoire des Arts et Métiers. A number of applied science and engineering schools and institutes were also established.

These centers, from which emanated cadres of professionals and skilled workers, reinforced and extended the reach of science and technology in districts across the city. At the École de Médecine and the École Normale Supérieure (ENS) on the rue d'Ulm in the central city's 5th arrondissement, the Ministry of Education and Fine Arts added buildings and laboratory space, including a wing where Louis Pasteur, by then internationally famous, expanded his facilities, But in 1888 he left to take possession of a new building in the eastern part of the 15th arrondissement, not far from the Latin Quarter, specially erected through government-held public subscription for his private research institute. Marie and Pierre Curie took over the rue d'Ulm space.

These institutions were more than marks on the urban map. They were nodes in a network for the exchange, increase, and application of scientific and technological information with a national and international reach.[63] Due to their placement they served a variety of constituencies and represented the city as a space where scientific and technological advances were channeled into jobs, industrial activity, and further urban improvements. Through their laboratories, they created the links between the *grandes écoles* and those schools devoted to applied technical education, and beyond the schools to the industrial sectors of the city as well. One example is the École Municipale de Physique et Chimie one street over from the ENS. On the Right Bank, in the 3rd arrondissement near the Conservatoire des Arts et Manufactures, the Conservatoire des Arts et Métiers in the rue Saint-Martin sported a new testing laboratory that served public and private clients.[64] At the far western end of the 15th arrondissement, abutting the chemical companies in Javel and near the first electrical companies founded in Paris, was the École Supérieure d'Électricité (known as Supélec), the product of private initiatives by industrialists looking to staff their firms and factories.[65] Its graduates and private testing institute served clients including the Compagnie Française Thomson-Houston, established in 1893, a sister company to Thomas Edison's General Electric.[66]

There were limits to this urban development, apparent for example in the restricted reach of public services such as telephone, electricity, and transportation. Despite the HBM legislation incentives, modern housing for workers continued to be very scarce because private developers saw no great profits to be made from construction or rents. There were plans in place to continue implementing modern improvements, but only as funding and political will permitted.[67]

EXPOSITIONS

As far as realized construction in Paris was concerned, however, change in these years made the science- and technology-based present more accessible to more French citizens than it had been in the previous era. Three vast universal expositions offered opportunities to fill some of the gaps in the city's infrastructure and provide material evidence and astounding symbols of continuous human improvement emanating from Paris.

The commissions charged with overseeing the 1878, 1889, and 1900 fairs saw them as opportunities to push urban development in new directions—and by the end of this period to introduce innovations that ushered in a highly mechanized vision of Paris.[68] Among these inventions were novel fiscal arrangements that wedded democratic republican government and industrial capitalist objectives. On a symbolic level, each exposition communicated a particular idea of the republic to the nation and the world: 1878 was intended as a vigorous statement that, eight years after defeat at the hands of the Germans and a civil war, France was back on track as a leading modern nation. In 1889 and 1900 boosterism was ratcheted up, as France claimed to be the country continuously inventing the future. In essence, Paris was reimagined as the nexus of a democratizing evolutionary process. If the 1889 exposition signaled the future made possible by the revolution of 1789's liberation of science and industry, that of 1900 proposed to be an accounting of science and industry's benefits to humanity so far and in years to come.[69] These gigantic events, filling ever-larger spaces in the city, were symbolic expressions of liberal democratic progress that enabled visitors to vicariously experience the future to which these symbols alluded.

The very process of constructing the fairs helped move the city into the future. The three expositions are identified with four districts whose development furthered the reach of science and technology in different parts of Paris: the areas surrounding the Trocadéro hill and the Champ de Mars; the Métropolitain; and the complex of streets and buildings that included the Champs-Élysées, the Grand and Petit Palais, the Pont Alexandre III, and the Left Bank. These projects in turn had a ripple effect on development in adjacent areas of the city and on Parisians and others visiting or doing business in the city.

Development of the Trocadéro hill provides a good example of how the Third Republic used expositions to extend imperial initiatives and turn them to its own ends. The lowering of the hill for the 1867 exposition was the beginning of development on what was then the edge of the city. In 1878, the government built the Palais du Trocadéro on the site; agricultural products were displayed there in that year's exposition. Designed by Gabriel Davioud, architect and inspector of city buildings for the City of Paris, and engineer Jules Bourdais, the Moorish-flavored, domed and turreted edifice sat above elegant gardens and a fountain designed by Alphand (figure 2.4). The space behind the hill and to the west had begun to be developed, but after the exposition of 1900, the wealthy 16th arrondissement and bourgeois Passy

FIGURE 2.4
Paris. *The Famous Trocadéro Palace from the End of the Seine Bridge,* constructed 1878, photo taken at the 1900 Exposition Universelle, Paris, France. Left half of stereotype by Underwood and Underwood. Library of Congress Prints and Photographs Division, Washington, D.C.

experienced development analogous to that occurring around the Champ de Mars.

The Trocadéro was conceived in such a way as to make it evident that modern technology could surpass the venerable monuments of traditional architecture, and that French builders could once again outdo the English. The dimensions of the main auditorium surpassed those of its English competitor, the Royal Albert Hall. The dome of the Trocadéro was some twenty-three feet higher than the dome of St. Peter's, and the flanking towers surpassed Notre Dame's tower by forty-five feet.[70] This was the beginning of the audacious humbling of the city's great religious monuments by the secular republic—a movement that would culminate eleven years later in the Eiffel Tower.

Across the river, the government tore down the decrepit housing that flanked the Champ de Mars to allow the 1889 exposition more room. After the 1900 exposition, the department of the Seine released the adjoining land to the city, which in turn sold it off to developers for construction of expensive residential and commercial neighborhoods. Here lighting, sewers, water, and wide streets—modern amenities already introduced to serve the expositions—were easily extended.[71]

Beginning in 1867, the railway station at the Champ de Mars was remodeled a number of times to accommodate changing circulation patterns and needs as expositions came and went, and as rail connections with the Métropolitain were built in the 1890s. The Gare d'Orsay, fitted to allow electrically powered engines, opened in time for the 1900 exposition with the express purpose of welcoming visitors to the fair from the south and west of France. With its rail connection to the Gare d'Austerlitz, it served as the point of debarkation for foreign delegations from the Austrian empire to the fair, the National Academy, and the Quai d'Orsay.[72]

As during the Second Empire, the expositions helped disseminate the culture of change even beyond their brief life spans. They left behind inventions geared to new urban experiences and needs, such as the Métropolitain and the complex linking the Champs-Élysées with the Left Bank. The Métro, first proposed in the 1840s, was finally constructed in advance of the 1900 exposition because organizers realized it would solve the problem of street-level congestion that would only increase as the event approached. No doubt competition with New York, Chicago, and London was also a factor in the decision.[73] Fulgence Bienvenüe supervised construction and designed the electrically powered system, which he planned to build in stages. The first line, opened in time for the 1900 exposition, ran on the east-west axis of the city from the Étoile, under the Champs-Élysées, to the Porte de Vincennes, where sporting events drew exposition visitors to stimulate development on the eastern edge of the city. Later lines also followed the routes of the streets above ground, echoing Haussmann's circulation plan for Paris. However, the left-wing Paris Municipal Council prevailed over the prefect of the Seine when it came to control over the design and construction. To prevent the railroads and railroad interests from entering into the heart of the city, the council insisted on narrow-gauge tracks, while suggesting placement of the projected lines to make inexpensive, rapid transportation available to Parisians of all social classes.

The specially designed entrances to the Métro aptly symbolized this subterranean break with Haussmann's urban plan for circulation. Aesthetically and technologically, architect Hector Guimard's fantastical green-iron vegetation and orange, insect-eyed electric lights signaled a shift in the culture of change. Their elongated tendrils stood out against the stone geometry of gaslit streets and opened the way to new, electrified experiences of space and time below ground.[74]

Exhibition planners returned to Haussmann's designs for Paris in the plan to join the lower end of the Champs-Élysées with the Left Bank. The idea of linking the two sides of the Seine at this point originated with Alfred Picard, who wanted to make the area the keystone of the 1900 exposition.[75] Planners incorporated a number of Eugène Hénard's ideas to integrate the exposition on the Left and Right Banks of the Seine, to enhance traffic circulation, and to provide an appropriately grand approach perpendicular to the Champs-Élysées. The project required intercepting the Champs-Élysées and constructing two large exhibition buildings (the Grand and Petit Palais) at this point, as well as a bridge over the Seine. It also made use of the boulevards, railway stations, sewers, and lighting systems already in place. Hénard's contribution, the Pont Alexandre III, was the organizing structure for a monumental urban ensemble; the bridge links the Champs-Élysées with the broad, open Esplanade des Invalides.[76] Visually, the bridge, with its innovative single-span design, was part of an urban perspective closed off by the dome of the Invalides. Functionally, it allowed traffic to flow across the Seine in both directions, from one part of the exposition and of the city to the other.[77]

Such celebrations of industrial might also inaugurated changes in the way the Parisian population and those contracted to build the exposition did business. Contracts with private firms that supplied materials (especially iron) required that they work within the protocols, standards, and methods set by the exposition planners. More dramatically, the republicans created funding arrangements that wedded the objectives of democratic government with liberal economic commitments to efficiency and profit-making investment in technological projects. Judging from the comments of Édouard Lockroy, the republicans were out to disprove Le Play's critique of expositions as wasteful economic endeavors by using them to demonstrate that democracy and industrial capitalism could work together for a better future.[78] Under government direction, funding for the expositions came from a combination of taxes, exhibit charges, and collaborations with business.[79] The collaboration between state and private enterprise was reflected in the choice of architects and engineers for the 1878 exposition. While both Gabriel Davioud and Jules Bourdais were state employees, the state commissioned an engineer from the private sector for the main entryway to the exposition: Gustave Eiffel's firm built the facade of the Palais de l'Industrie, designed by Léopold Hardy.[80]

Besides altering the physical city, the expositions also allowed republican elites to reify and vivify their visions of the new industrial order. No monument of the Third Republic symbolized these open-ended possibilities better than the Eiffel Tower. Eiffel himself saw the tower as a thing of beauty, its asymptotic curves the material equivalents of geometrical laws and the laws of physics.[81] Lockroy chose to interpret the form in more political terms, writing that the Eiffel Tower "summarizes the industrial grandeur and power of the present. Her immense spire, buried in the clouds, has a symbolic quality; it is the image of progress as we conceive it today: an unending spiral where humanity gravitates in its eternal ascension."[82] And no single project could more dramatically move French society to enact the creation of liberal democracy. For Lockroy, the tower engaged millions of visitors from Paris, the nation, and the world in an experience that touched their imaginations and sensibilities, inspiring them, he asserted, with feelings of controlled ascent and comradely support that they would bring back to earth with them. Added to this "experiential learning" was the symbiotic relationship the Tower project inaugurated between public and private enterprise. Eiffel and the French government committed to sharing the costs and risks of construction.[83] Looking ahead to the profit-making potential of the giant edifice, Eiffel agreed to repay the state's investment in exchange for the right to the concession for the tower, which he would turn over to the state at a prearranged date.[84]

The great iron edifice stood as a gateway to the 1889 exposition, where as part of a vast ensemble of iron structures it served the function of a giant triumphal arch. Through it one entered a U-shaped arena in which the placement of the Galerie des Machines and the palaces of the fine and liberal arts constituted a schema of industrial society, where fundamental production processes supported and benefited from the products of intellectual and artistic labor. The entire exposition, Lockroy argued, like the revolution it

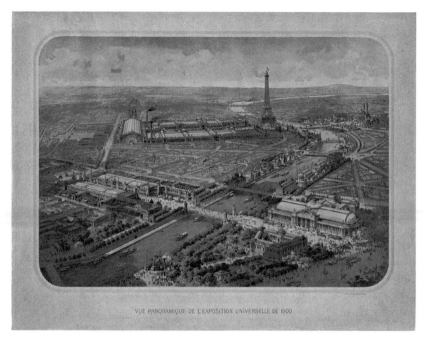

FIGURE 2.5
Paris. Aerial view plan of 1900 Paris Exposition Universelle Internationale. Library of Congress Prints and Photographs Division, Washington, D.C.

celebrated, was "a glorious event in our history … the point of departure for the entire world of a new era."[85]

As if to show what that new era turned out to be, the 1900 exposition was a gorgeously clothed and larger version of 1889, reworked to emphasize the promises of growing consumer markets and the world empire in which Paris participated (figure 2.5). The exposition's theme was "An Accounting of the Century," as Picard's introduction in the official catalog is careful to explain.[86] Through overwhelming amounts of evidence on display, it attempted to prove that the outcome of scientific and technological innovation had been and would continue to be a better life for everyone in the world. This message was communicated to visitors as they moved from the painted and bejeweled entry arch crowned by a fashionably dressed female figure toward the Champs-Élysées across the Pont Alexandre III and onto the Esplanade des Invalides. Flanking their path were the Grand Palais and the Petit Palais, housing fine arts exhibits. Art Nouveau reigned as the style of middle-class domestic interiors, and of decorative arts objects from armoires, desks, and electric lamps to innovatively processed glassware.[87]

If visitors chose to enter from the Trocadéro across the Seine, they passed through gardens where the colonial holdings of France took up one side and those of other world powers the other. The Trocadéro Palace itself, used for numerous international scientific and technological congresses at the fair, might be considered the cerebral cortex of that great Parisian brain Zola had conjured up almost twenty years earlier. Crossing the Seine, visitors passed under the Eiffel Tower to face the Château d'Eau and behind it the stuccoed Palais de l'Électricité, designed by Hénard, stretched across the Champ de Mars. The Palais was adorned with electric lights. On exhibit there were "all the applications of electricity, telephonic systems and all recent electrical inventions," while hidden from view were the dynamos that supplied this newly tamed power source to the entire exposition, through "invisible wires and powerful motors."[88] The complicated negotiation of past, present, and future that electricity posed for fair organizers had perhaps no better example than the display at the center of the Château d'Eau. Here, colored electric lights played on a giant Louis XV-style waterfall fed by water electrically pumped from the Seine that exited into a pool decorated with a thirty-foot-tall Beaux-Arts sculpture group with the allegorical title "Humanity Guided by Progress Advances Toward the Future."

To look at the expositions sequentially, the examples of the Eiffel Tower, Château d'Eau, and Palais de l'Électricité suggest that first the 1889 and then the 1900 exposition gave elites the opportunity (or, perhaps, even forced them) to construct time and space in more fluid ways. Like the fairs mounted during the Second Empire, each was an attempt at bundling human products, activities, and cultures into intellectually coherent, political narratives that privileged industrial society, and in the case of the 1889 and 1900 exhibitions, the republic's sense of its historic mission. This organizing was carried out through standardized rules and regulations for submission and display, but it was materialized in the catalogs and especially in the site plans of the expositions themselves.

Progressive time, in Lockroy's view, was made manifest in the nationally organized exhibits at the 1889 exposition. On view were the material fruits of Enlightenment ideology that had set France at the head of Western progressive movements. If the 1889 extravaganza showed what national adherence to science and technology could accomplish, it also acknowledged the great revolution of 1789 as inaugurating the economic and political liberty necessary for Enlightenment intellectual freedom to function in society's interests.[89] In the context of international competition, the present on view in Paris was but a spur to further improvement for each nation under the aegis of French republicanism.

In 1900, progressive change and complementary social arrangements supposed to drive it jostled for logical expression in the exposition's plan. Picard devised a novel organization that downplayed the old national competitions in favor of broad classifications corresponding to the economic and social categories around which industrial societies were becoming internationally structured.[90] The fact that the fair's ornate entry arch was topped by a statue of La Parisienne (which some said symbolized Peace), while the Seine was filled with warships from the participating nations, speaks of the contradictions and hopes alive in Picard's plan. In an exposition that both measured the century's accomplishments and wished to set a course for the future, organizers hoped emphasizing international perspectives on the benefits of science and technology might deflect current hostile competition into future peaceful ends.[91] Thus, countries competed within palaces such as those on Instruction and Education, Chemical Industries, and Civil Engineering. On the Trocadéro hill, private enterprises and agencies doing business in (or with) the colonies mounted exhibits of everything from Parisian-built hydraulic and transport systems to reports, charts, and graphs on plantation administration and schools.[92] Following official guidelines for all participants, French exhibitors played down the race for world domination by showing how circulation of industrial goods and ideas under the aegis of the mother country was a force for mutual, if unequal improvement.[93]

The narrative of industrial progress was alive at the fairs, showcased in exhibits that provided popular experiences of what might lie in store for urban consumers in the future. Such experiences included the moving sidewalk that carried visitors along the Seine between the entry and the Champ de Mars; the electric automobiles manufactured in the city that helped distribute hundreds of place settings for the visiting mayors' banquet and available for private rental; the lifts for the Eiffel Tower; the aeronautic demonstrations across the city's skies; and the electrical powering of numerous domestic, urban, and industrial technologies on view. But Paris was very much present in exposition participants' plans for shaping the future path of several pressing issues: public health, meteorology and aeronautics, and social economy. In the case of public health, Paris took the lead in 1900 when the international commission created by the medical congresses there adopted the city's methods for keeping track of mortality rates and causes.[94] Using the system devised by physician Jacques Bertillon, head of Paris's municipal department of vital statistics, as a model, they created an international standard for nosology and data analysis.

In 1900 scientists from Paris-based institutions also dominated the international congresses devoted to meteorology and aeronautics. The meteorologists focused on aerial achievements that would increase knowledge of earth's atmosphere and might eventually allow humans to escape it. Meanwhile, as part of the aeronautics division, the French minister of war sponsored a private tour for attendees showing off the army's nearby balloon factory. With Germany's scientists and military very much in mind, organizers of the aeronautic exhibition on the Champ de Mars featured a variety of French inventions, including the Avion flying machine designed by Clément Ader. A batlike construction based on Marey's studies of birds, it was powered by a steam engine and offered a promising—if as yet unproven—approach to conquering the skies. A seven-mile dirigible race, between Vincennes and the Eiffel Tower, was planned, and the Aéro-club de France sponsored demonstrations and competitions that filled the skies of Paris with balloons and newly developed gasoline-powered steerable dirigibles. Contests pitted nations against one another in races to go higher, farther, and faster with greater accuracy.

A number of scientists at the meteorological and aeronautical congresses worried that by working to achieve these goals, they opened possibilities for destroying the very society they had participated in building. The distinguished astronomer Jules Janssen, head of the French astrophysical observatory and president of both congresses, introduced the disturbing specter of what progress in aeronautics on view at the exposition augured for the international industrial order. Rather than focusing on the possibilities opened for mail delivery or rapid commercial travel between distant cities, the American journal *Science* reported, "M. Janssen predicted that the nation which first learned to navigate the air would become supreme, for while the ocean ... has its boundaries, the atmosphere has none. What then ...

will become of national frontiers when the aerial fleets can cross them with impunity?"[95] German dirigibles, floating in to bomb the city in 1915, would soon prove Janssen's fears correct. Mastery of the air threatened the destruction of the modern city that republican progress had wrought.[96]

The organizers of the social economy exhibits in 1889 and 1900 were more optimistic and literally more down-to-earth in their focus on using science and technology to solve social problems. The exhibits and projects on view included both direct and indirect approaches to integrating workers into modern society, and proffered scientific solutions to achieve this end. Not only did they distill their existence into scientific data, statistical charts, and photographic evidence; but among these solutions were proposals and plans for inexpensive urban housing that was connected to sewers and other infrastructural improvements. Government officials and ministries as well as private societies were involved in mounting these exhibits, which in turn generated congresses, new organizations, committees, and international exchanges of research and individuals all focused on bringing the culture of change to bear on the lives of working-class people.[97]

While the social economy exhibits were not entirely devoted to Paris, the capital city held a primary place in the pavilions as both the subject of and home to the agencies and groups that organized the displays. Likewise, the congress held in the Pavillon d'Économie Sociale in June 1900 turned Paris into a major center for the international organization, coordination, and dissemination of information on social economy, setting into motion a series of historically important institutional developments centered in the city.

As mentioned above, enthusiasm among followers of Le Play and conservative republicans—most notably Jules Simon and Jules Siegfried—led to the formation of the Musée Social in 1894. The Musée took a special interest in urban housing, working to draw up legislation to encourage investment in low-cost housing construction, and establishing sections on urban hygiene and urban housing.[98] Along with the Société Française des Habitations à Bon Marché (1894), and later the Société Française des Urbanistes (1911), the Musée Social constituted a novel and important nexus for architects, politicians, sociologists, and reformers to develop proposals for housing design and social hygiene and legislation to encourage investors to construct the buildings.[99] In combination with its broader vision of the modern city and the nation as social systems necessary for the operation of the modern industrial order, the Musée Social's support for urban housing research would make it one of the cradles of French urbanism in the decade before World War I.

Eugène Hénard was perhaps the most important architect among the men who met at the Musée Social in these years. He was also a charter member of the Société Française des Urbanistes and contributed an important paper to the first meeting of the International Society of Urbanists in London in 1911. While his architectural impact on Paris, as discussed above, was more significant than is usually acknowledged, Hénard is arguably more important as a quintessential urbanist. In essence, he helped found a new profession that saw the city as a rational whole, a system that could be scientifically planned, designed, and built to accommodate new industries, new power sources, and new modes of transport. By the first decade of the twentieth century, he had proposed a universal plan for the city of the future. Like Jules Verne in *Paris in the Twentieth Century*, Hénard identified nascent industries and extrapolated a picture of their technologies as dominant animators of the future city. Yet unlike Verne forty years earlier, as a practicing architect Hénard could make use of statistics and research to forge a convincing, plausible plan. He replaced the old hodgepodge of different systems (roads, sewers, lights, transport, buildings made of stone and wood) with one rationalized, mechanized, and integrated system of circulation, including the automobile, the airplane, and electricity, which extended into the air as well as below and above ground.[100]

In seeming contradiction to the focus on the urban future in Paris, a number of exhibits and concessions took a "forward to the past" approach toward historical continuity, congenial to Lockroy's and Picard's progressive programs for 1889 and 1900. Rather than providing escapes from the present, they evoked the history of urban life in very modern terms. Connections between present and past in the city were evident in Parisian-themed popular amusements. In 1889, visitors to the Bastille attraction rode on wheeled wooden horses, which sped them up and down on a roller coaster track through a mock reconstruction of the fortress where the Revolution—here defanged by arcade machinery—had begun. The messy deaths, popular uprisings, and profound emotions once brought to mind by the Bastille were now smoothed into safely historicized thrills of technologically driven motion. In 1900 visitors could take the new Métro to the popular Old Paris concession on the Right Bank. There they wandered through plaster-and-paint reconstructions of medieval buildings on cobblestone streets bordered by artisans' shops and restaurants, but unencumbered by churches or prelates. In this mock city, the present was always "present."[101] Electrically illuminated at night and free from such premodern historical realities as rotting sewage, poverty, and religious and social strife, Old Paris seems to have idealized the city's past, attributing to it the standards of modern republican Paris, whose roots could be found there and whose "real" spires—Notre Dame, the Conciergerie, and the Eiffel Tower—were clearly visible.

The 1900 City of Paris pavilion itself offered an official variant of this collapsing of past and present with the future. Built of wood rather than iron and glass, in a style reminiscent of the old Hôtel de Ville, the pavilion sported Paris's medieval coat of arms and emblems, recalling the city's ancient trades and corporations. Inside, their supposed contemporary incarnations, the city's newly founded departments of Public Roads, of Light, Water and Drains, and those of the Quarries and new Metropolitan Railroad, mounted statistical charts and photographic evidence of the latest infrastructural improvements to the city. On the floor above, the major educational institutions and libraries of the city had organized exhibits.[102] A cinematograph projected educational movies of these departments at work, adding to the impression that these organized and organizing municipal activities stretched from the present into the future.[103]

By comparison with these efforts, Charles Garnier's large exhibit on the History of Human Habitation at the 1889 exposition offered a more abstract approach. Rather than including Parisian dwellings, his vision of change could be applied to any contemporary example. As Garnier explained in the book that accompanied the exhibit, he imposed a modern system of chronological development on a panoply of ideal types, starting with the shepherd's hut and ending with the modern city house. In creating this taxonomy, he seems to have followed a procedure that combined analytical approaches well established among his colleagues at the École d'Architecture with those of professors at the Muséum, adding a nod to ethnography as well. It is a method that John Pickstone has called the "analytical/comparative or museological/diagnostic," linking social and cognitive forms— in Garnier's case, abstracting out and comparing select characteristics of domestic living spaces.[104] Although Garnier sidestepped the more politically charged issue of comparing the design of worker housing across historic cultures, his exhibit nevertheless focused on a topic of intense interest among contemporaries.

There is no doubt that expositions under the Third Republic materially advanced the Haussmannian plan for the city. These mammoth fairs did contribute to Paris's becoming a more technologically organized space through urban rebuilding and construction of the Métro and rail links, and they increased the connections between the city, the hinterlands, and the world beyond France's borders, including the expanded empire. Yet the republic's expositions took routes politically distinct from those of the Second Empire. Following Jules Ferry's hopes, late-nineteenth-century expositions in Paris extended the international and institutional reach of liberal science and technology into future-oriented projects that attempted to reign in or steer capitalist agendas into democratic objectives. The Paris Métro, although designed by a *polytechnicien*, was a project of a left-leaning city council using the exposition as an opportunity to keep railroad interests at bay. Moreover, the expositions generated French-led international organizations of rising professionals from the emerging bureaucracies of industrial nations. Through these networks for collecting, exchanging, and disseminating knowledge about science and technology, urban society, and urban planning, the French led the way in setting standards and regulating cooperative arrangements between governments, private industry, and commercial establishments. Metaphorically these organizations might be considered expressions of Zola's vision of Paris as a great brain where intellectual liberty was a force for peace. But the opposite was also true of the expositions: In turning the city into a stage for demonstrating electricity, the automobile, and airpower at the fair, republican elites helped open the door to a wholly different level of industrial existence, one characterized by greater interconnection at a faster pace, yet threatened by new forms of disorder.

To step back a bit, we can see that these expositions also demonstrated a palpable shift in perceptions of time in the city and in conceptions of history itself. Organizers and participants in committees, congresses, and commissions redefined Paris under republican auspices as a place where elites increasingly evaluated scientific and technological achievements more in terms of their organizing power and effects on progress than in terms of their actual contribution to the material improvement of city life. At the same time, in the process of creating organizational structures, they turned the products of science and technology into artifacts of an ongoing history.

MUSEUMS

In comparison with the Second Empire, which isolated mementos of the Parisian past in a municipal museum set amid a sea of new construction, Paris in the early Third Republic was what can only be described as a city "museified" on scientific and technological terms. Spurred by the staggering growth of the tourism industry as a result of the expositions, Paris emerged as a vast showcase, where science and technology linked the past with the industrial present. Simply as the result of the aesthetic character imposed on it by the building codes, the city became a panorama integrating past and present structures into one continuous experience. Moreover, with encouragement from republican ministers bent on making Paris the intellectual center of its new empire and the civilized world, a new breed of specialists worked to make the growing number of Parisian museums into important centers for collecting, organizing, and disseminating information about the natural and human world. An evolutionary schema set forth Paris as the great organizer of industrial progress.

The museification of the city was accomplished partly through the officially mandated system of proportions for street width and building and story height. In conjunction with the official preference for neoclassical, neobaroque, and Beaux-Arts styles, constructed in sandstone and with limited color schemes, these formulae created modern architectural frames aesthetically compatible with older buildings.

To some, including the artists and writers who signed the infamous petition against it, the Eiffel Tower seemed a shocking anomaly in this Parisian display. Visually, culturally, and even morally it represented a break between a preindustrial past and the modern capitalist present.[105] But supporters saw the tower—through the lens of republican reform—as integral to the modern city, arguing that this iron skeleton laid bare the same rational principles and values at work, though hidden, in great historic monuments such as the Arc de Triomphe and Notre Dame. This particular vision of Paris emphasized its role as a museum of technology, the city as the site of an ongoing engineering tradition.[106]

Guidebooks both reflected and helped create the panoramic conception of Paris as a place filled with scientific and technological points of interest.

While these guidebooks provide an outsiders' view of how museums in Paris were shaping notions of the past and present, their need to be "au courant" makes them a good means for identifying changes going on within the institutions themselves. From them we can tell that some of the most important museums contained scientific and technological collections, some sought scientific and technological information about the objects in their collections, and that most of their curators wanted to organize their collections using classification systems that reflected natural laws governing change. Hence, the Musée de la Ville de Paris, reestablished and reconstituted in the decade after the Commune, divided its collection by chronological periods in the city's history, ending with the Revolution. A nineteenth-century section was planned. The original mission of the Musée des Arts Décoratifs, founded in 1882, included the classification, by period and culture, of manufacturing processes and materials used to produce objects such as ceramics, textiles, glass, and furniture.[107] At the Conservatoire National des Arts et Métiers, curators during this period did extensive research and reinstalled the galleries, publishing a detailed catalog of the mechanical arts collections in 1911; this was designed as a visitors' guide through the history of this important branch of nineteenth-century technology.[108] The Muséum d'Histoire Naturelle (renamed the Muséum National d'Histoire Naturelle under the Third Republic), including the gallery of anthropology at the museum in the Jardin des Plantes and the Musée d'Ethnographie at the Trocadéro, all counted technological artifacts (useful objects, tools, etc.) as important parts of their collections; at each institution the concern was to develop and apply scientific schemas that privileged differences in time and place among like types of objects so as to rank them in a progressive order with Western civilization at the forefront.[109]

Officials in the Ministry of Public Education and Fine Arts, such as Ferry and Antonin Proust, encouraged museums to pursue these goals through legislation.[110] In addition, societies (whether designated private or of public utility) were important forces for mobilizing museum programs and founding specialized museums to pursue scientific agendas. Among the most active advocates of museums were the Union Central des Beaux-Arts Appliqués à l'Industrie, the Société du Musée des Arts Décoratifs, the Société d'Anthropologie, and the Société d'Ethnographie. Organizations often competed with one another for control over emerging areas of specialization and the associated artifact collections.[111]

The government and specialists did agree on the main reasons for supporting museums—all of which were replete with republican historical consciousness. Highest on the list were national prestige, economic benefits, and usefulness to science.[112] Both groups also succumbed to the opportunity to increase the number of national treasures by purchasing large quantities of artifacts and even entire exhibits on view at the international expositions.[113] In the highly competitive climate of late-nineteenth-century industrial societies (Berlin, Chicago, and London already had impressive sets of new museums), institutional acquisitions promised to place France at the head of modern historical forces.

The history of the Musée d'Ethnographie du Trocadéro under its first director, Ernest Hamy, provides a good example of how specialists, charged with classifying collections, added goals that reflected their own research and disciplinary objectives.[114] Hamy started the ethnographic collection as a curator of the anthropology collections at the Muséum d'Histoire Naturelle, where he sorted out and attempted to classify what he felt were ethnographic rather than anthropological artifacts. Hamy saw the announcement of the 1878 exposition as an opportunity to add ethnographic materials to those he had already culled from the anthropological holdings. It also inspired him to begin lobbying the government for a separate building that was well placed in the city and architecturally suited to exhibiting the collections in a coherent way. With the support of a commission that included Jules Ferry, the result was an ethnographic museum, officially founded in the east wing of the Trocadéro Palace in 1879 (it was a predecessor to the acclaimed Musée de l'Homme, established in a new building on the same site in 1937). Here Hamy could consolidate materials previously dispersed in a number of museums, sort through artifacts obtained from the expositions, and analyze, classify, and display them for research, public education, and profit.[115] The 1889 exposition allowed him to fill out the collections with objects from the new South Asian and African colonies, as well as from North Africa, China, Japan, and Amerindian cultures. The 1900 exposition served the same purpose, while also helping to highlight the museum's collections in the Trocadéro wing, located just above the French colonial exhibits.

Hamy's definition of ethnography was broad, and he made certain to couch the new field of study in terms of the contributions it made to understanding human history and human activity generally, and to the other disciplines with which he was competing. The "new" science of ethnography was really the binding agent in a growing knowledge network.[116] Taking his cue from the historian Hippolyte Taine (1828–1893) and from Scandinavian ethnographers, Hamy founded his classification system on the premise that artifacts were expressions of a common human psychology and the products of common human needs. Organizing them by culture and date and focusing on certain traits or physical characteristics of objects allowed for historical and evolutionary comparisons. This approach had the practical advantage of allowing him to include artifacts from both colonial and European cultures within a universal framework. His classification system began with the physical types or races, and then moved to basic human necessities for survival, up a chain of increasingly elaborate activities: food, defense, dwellings and ways of living, furnishings, means of communication, industry and commerce, arts and sciences, religion, and social life.[117]

It also might be argued that Hamy's system was a partial realization of that story of human progress found in Condorcet's famous *Sketch for a Historical Picture of the Progress of the Human Mind,* for he displayed peoples within a common framework based on selective empirical observations. His developmental comparisons were based on the Enlightenment ideals and values prized by republican elites: the primacy of the individual, the family as foundation stone of society, the centrality of work and invention in human progress, and the irrational character of religious beliefs and practices. In a very concentrated way, Hamy's museum was a place where the republicans' vision of a future order was brought around full circle to encompass the past. In this sense, it was fully in tune with other Parisian institutions that constructed narratives of the city's evolutionary change using the language of science and technology.

Republican elites transformed the urban past to conform to the direction in which they wished to turn French society. And they made the past accessible to the public. Through their extension of Haussmann's plan which had coherently wedded old to new construction, they had museified the city, representing its democratic, industrial present as implicit in its past. Something analogous to this abstraction of the city also occurred within museums—especially those dedicated to scientific and technological subjects. By 1914, curators in Parisian museums had made the city an international center for the invention and dissemination of classification systems that integrated world cultures into an evolutionary representation of the world's past, providing an explanatory framework into which all future discoveries and productions of civilization could be fit.

PART III   CONCLUSION

Between 1852 and 1914, Paris developed into a modern center of industrial society through the efforts of elites who combined institution building with a belief in the power of science and technology to organize positive change. The mechanisms they used—urban rebuilding and development, universal expositions, and museums—were part of a culture they invented to turn Paris, and through it the French nation, into an organized, powerful society. It is clear that the work of Napoleon III and Baron Haussmann laid the foundations for modernity (as historians have long recognized), but their engagement of industrial capital and urban rebuilding was only the beginning of the process. The elites of the Third Republic eliminated the major obstacles to extensive change, founded new educational institutions, and deployed a range of public and private organizations, government agencies, and institutions to extend the reach of science and technology and democratize industrial capitalism. Also, as we have seen, the rebuilding of the city does not in itself explain how this profound change in culture, in social existence, and in conceptions of time and space came about. The elites' commitment to universal expositions (cities within the city) and to museums was an essential factor on all these counts.

The logic of Haussmann's plan and Napoleon III's vision was carried through in the Third Republic's development of the urban fabric of Paris. Elites did create a very different kind of city by century's end. A sanitary city—a city of engineered streets, with interconnected systems of light, heat, and water—was realized for the middle classes, even if it was not extended into the working-class periphery. The Métropolitain fit into the general desire for a rapid, mechanized transportation network. It seemed that the plans for the future of the city had come to fruition.

But Parisian elites were to find that the very modern order they created in Paris had unexpected consequences. By 1900 the mid-century idea of the modern city based on steam and railroad industries had played out to its logical conclusion. The advent of electricity, the electric- and gasoline-driven automobile, and the airplane, as well as steel, were the signifiers of a new, more exciting, and potentially dangerous era, freed from the old earthbound constraints. It was Hénard who saw that Paris was old, and needed to be rebuilt to accommodate the coming ways of life associated with these new technologies, forms of industrial organization, and reconfigured social spaces. He began to design its future on paper, but the city's leaders lacked the political will and economic resources to begin anew.

By 1914, nineteenth-century modern Paris had reached its limits. The interwar period saw the plan for the Métro realized, a few boulevards completed, a few museums added in buildings constructed for the Coloniale and Arts Décoratifs expositions, and the Trocadéro Palace replaced by the Palais de Chaillot. But these accomplishments simply continued the old solutions rather than addressing how to revise the culture of change for new technological, social, and urban conditions. Although the emperor's and the Third Republicans' plans for inexpensive housing and limited amenities remained stalled, their solutions were in any case inadequate and outdated for the traffic-jammed city and its industrial peripheries, where automobile factories and chemical plants were surrounded by immense slums occupied by their workers. In the absence of funding and political backing to redesign and rebuild during these hard decades, architects and planners like Le Corbusier turned to their drawing tables and writing pads to imagine a different Paris of the future. Not until after World War II, during the years of the *trente glorieuses* under the Fourth and Fifth Republics, would the dynamics of capitalism once again take hold of Paris. Then, a generation of *dirigiste* elites would appropriate the culture of change to try to reconstruct the city in accordance with another set of industrial ideals.

NOTES

1. Jules Verne, *Paris in the Twentieth Century* (New York: Random House, 1996).

2. Ibid., 7.

3. For studies that discuss continuities and breaks in the development of Paris, see Pierre Lavedan, *Histoire de l'urbanisme à Paris* (Paris, 1975); Anthony Sutcliffe, *The Autumn of Central Paris* (Montreal: McGill-Queen's University Press, 1971); Norma Evenson, *Paris: A Century of Change, 1878–1978* (New Haven: Yale University Press, 1979); Colin Jones, *Paris: Biography of a City* (London: Penguin, 2006); Johannes Willms, *Paris, Capital of Europe: From the Revolution to the Belle Epoque*

(New York: Holmes and Meier, 1997). For general works on France that treat continuities and divergences between the two regimes, see Alain Plessis, *The Rise and Fall of the Second Empire, 1852–1871,* trans. Jonathan Mandelbaum, Cambridge History of Modern France, 3 (Cambridge: Cambridge University Press, 1999), 39, 43; and Jean-Marie Mayeur and Madeleine Rebérioux, *The Third Republic from Its Origins to the Great War, 1871–1914,* trans. J. R. Foster (Cambridge: Cambridge University Press, 1987), 88, 109, 111, on continuities in education reform, for example. The classic study on the Second Empire is David H. Pinkney, *Napoleon III and the Rebuilding of Paris* (Princeton: Princeton University Press, 1958); but also see David Harvey, *Paris, Capital of Modernity* (New York: Routledge, 2006); and David P. Jordan, *Transforming Paris: The Life and Labors of Baron Haussmann* (New York: Free Press, 1995).

4. Christian Topalov, "Les 'réformateurs' et leurs réseaux: Enjeux d'un objet de recherche," in Topalov, ed., *Laboratoires du nouveau siècle: La nébuleuse réformatrice et ses réseaux en France, 1880–1914,* Civilisations et sociétés, 98 (Paris: École des Hautes Études en Sciences Sociales, 1999), 11–48; Daniel T. Rodgers, *Atlantic Crossings: Social Politics in a Progressive Age* (Cambridge, MA: Belknap Press of Harvard University Press, 1998). The Topalov book is reviewed by Michael A. Osborne in *Isis* 92, no. 3 (September 2001), 614–615, http://www.jstor.org/stable/3080773.

5. See the speeches on December 10, 1850, September 11, 1851, and October 9, 1852, in *Discours et messages de Louis-Napoléon Bonaparte depuis son retour en France jusqu'au 2 décembre 1852* (Paris, 1853), 136–138, 150–151, and 240–243, respectively.

6. Theodore Zeldin, *Émile Ollivier and the Liberal Empire of Napoleon III* (Oxford: Clarendon Press, 1963). One architectural historian who links the philosopher's ideas to economist Michel Chevalier and Napoleon III's interest in urban planning and expositions is Spyros Papapetros, "Paris Organique—Paris Critique: Urbanism, Spectacle and the Saint-Simonians," *Iconomania: Studies in Visual Culture,* www.humnet.ucla.edu/humnet/arthist/Icono/papapetros/simonian.htm, 25–26 and notes 66–67.

On *polytechniciens* and Saint-Simonian ideas, see Francine Masson, Antoine Picon, and Béatrice de Andia, *Le Paris des polytechniciens: Des ingénieurs dans la ville, 1794–1994* (Paris: Délégation à l'Action Artistique de la Ville de Paris, 1994); Antoine Picon, *L'invention de l'ingénieur moderne: L'École des Ponts et Chaussées 1747–1851* (Paris: Presses de l'École Nationale des Ponts et Chaussées, 1992); and César Daly, "Toast aux ingénieurs," *Revue générale de l'architecture et des travaux publics* 34 (1877): 101. Also see Bernard Landau, "La fabrication des rues de Paris au XIXe siècle: Un territoire d'innovation technique et politique," *Annales de la recherche urbaine* 57–58 (December 1992–March 1993): 23–46. References here to this article are based on an unpaginated Web version: http://www2.urbanisme.equipement.gouv.fr/cdu/datas/annales/landau.htm.

For the general ideas of the Saint-Simonians in relation to architecture and urban visions, see Ann Lorenz Van Zanten, "The Palace and the Temple: Two Utopian Architectural Visions of the 1830s," *Art History* 2, no. 2 (June 1979): 179–200; and Anthony Vidler, who gives an account of the ideas of Saint-Simon and his followers in "The Scenes of the Street," in Stanford Anderson, ed., *On Streets* (Cambridge, MA: MIT Press, 1977), 28–111, especially 58–60. For general discussions of the philosopher's ideas, see Frank Edward Manuel, *The New World of Henri Saint-Simon* (Cambridge, MA: Harvard University Press, 1956); and *Henri Saint-Simon (1760–1825): Selected Writings on Science, Industry, and Social Organisation*, trans. and ed. Keith Taylor (London: Croom Helm, 1975).

7. Jordan, *Transforming Paris*. For Haussmann's own account, see Georges Eugène Haussmann, *Mémoires du Baron Haussmann* (Paris: Victor-Havard, 1890), 3 vols., republished as *Mémoires: Édition Intégrale* (Paris: Seuil, 2000).

8. See Pinkney, *Napoleon III and the Rebuilding of Paris*; and Jordan, *Transforming Paris*. For more recent studies on continuities and breaks in the development of Paris, see Lavedan, *Histoire de l'urbanisme à Paris*; and David Van Zanten, *Building Paris: Architectural Institutions and the Transformation of the French Capital, 1830–1870* (Cambridge: Cambridge University Press, 1994).

9. At the 1867 exposition, which he directed, Le Play showcased plans for low-cost, sanitary urban housing in the form of *cités ouvrières*. These projects proposed modern hygiene as the solution to class conflict and the moral degradation of the poor. Just as dramatic and as crucial for different reasons, the Ministry of Public Instruction's exhibit showcased Duruy's plans for the future of French higher education, with a heavy emphasis on science. Although he could not garner the funds to build what was needed to advance French science and technology, he exposed present conditions that put France at a disadvantage compared with Germany by featuring a volatile report on the state of French research and science education, with contributions by Louis Pasteur and Claude Bernard. He also propagandized his proposals in the hope of winning public support for his particular vision of the future.

For a discussion of the relationships among these men and the differences between the social objectives and means (including expositions) of the Imperial *équipe* and that of Musée Social founders, see Topalov, "Les 'réformateurs' et leurs réseaux." Paul Rabinow, *French Modern: Norms and Forms of the Social Environment* (Cambridge, MA: MIT Press, 1989), traces the trajectory of the development from Second Empire and Third Republic expositions to the Musée Social. Janet R. Horne, *A Social Laboratory for Modern France: The Musée Social and the Rise of the Welfare State* (Durham: Duke University Press, 2002), discusses the role of expositions in preparing the way for the Musée, founded in 1894.

For Duruy's efforts related to the 1867 exposition, see Patrick Harrigan, *Lycéens et collégiens sous le Second Empire: Étude statistique sur les fonctions sociales de l'enseignement secondaire public d'après l'enquête de Victor Duruy (1864–1865)* (Paris: Éditions de la Maison des Sciences de l'Homme, 1979 [distributed by University Microfilms International, Ann Arbor]); and the series *Recueil de rapports sur les progrès des lettres et des sciences en France* (Paris: Imprimerie Impériale, 1867), in which Claude Bernard published the *Rapport sur les progrès et la marche de la physiologie générale en France, Publication faite sous les auspices du Ministère de l'instruction publique.* Also see Sandra Horvath-Peterson, *Victor Duruy and French Education: Liberal Reform in the Second Empire* (Baton Rouge: Louisiana State University Press, 1984), 177–179.

For references to the expositions and their role in the history of French science during these two regimes, see J. Franklin Jameson, "Introductory Notice," in Victor Duruy, *A History of France*, trans. M. Carey (New York: Thomas Crowell, 1889), xi; Robert Fox, "France in Perspective: Education, Innovation, and Performance in the French Electrical Industry, 1880–1914," in Robert Fox and Anna Guagnini, eds., *Education, Technology and Industrial Performance in Europe 1850–1939* (Cambridge: Cambridge University Press, 1993), 203, 210, 222, n. 6, and 224, n. 32; and "France" in Fritz Ringer, *Education and Society in Modern Europe* (Bloomington: Indiana University Press, 1979), 113–156.

10. See Van Zanten, *Building Paris*, 157–159, and n. 6; and the entry "Viollet-le-Duc, Eugène-Emmanuel" in Jonathan Woodham, *A Dictionary of Modern Design* (Oxford: Oxford University Press, 2004). *Oxford Reference Online.* Oxford University Press, 21 December 2008, http://www.oxfordreference.com.

11. See note 10 above; see also the surveys published by the Commission des Travaux Historiques (1865) and the municipal Bureaux des Travaux Historiques (1860) in the series "Histoire générale de Paris; collection de documents fondée, avec l'approbation de l'empéreur, par m. le baron Haussmann et publiée sous les auspices du Conseil municipal" (Paris: Imprimérie Impériale, 1866).

12. Among the many articles and books on the Third Republic, see in particular D. R. Watson, "The Politics of Educational Reform in France during the Third Republic 1900–1940," *Past and Present* 34 (1966): 81–99; M. J. Burrows, "Education and the Third Republic," *Historical Journal* 28, no. 1 (1985): 249–260; and Sanford Elwitt, *The Third Republic Defended: Bourgeois Reform in France, 1880–1914* (Baton Rouge: Louisiana State University Press, 1986).

13. Useful histories of Paris during this period include Jones, *Paris: Biography of a City*, with an emphasis on continuities, 331–333; and Willms, *Paris, Capital of Europe.* Also useful are André Castelot, *Le grand siècle de Paris, de la prise de la Bastille à l'effondrement de la Commune* (Paris: Amiot-Dumont, 1955); Christophe Charle, *Paris fin de siècle: Culture et politique* (Paris: Éditions du Seuil, 1998); Nobuhito Nagaï, *Les conseilleurs municipaux de Paris sous la IIIe République (1871–1914)* (Paris: Publications de la Sorbonne, 2002); David Harvey, *Consciousness and the Urban Experience: Studies in the History and Theory of Capitalist Urbanization* (Baltimore: Johns Hopkins

University Press, 1985); and Harvey, *Paris, Capital of Modernity*, xii–xiii, 24–25. Although Harvey refers to the important role of scientific and technological experts, he concentrates on the economic restructuring of time and space, and its attendant effects on growing consciousness of the urban condition. Rabinow's *French Modern* traces the exportation of this French version of modernity from its invention in Paris in the Second Empire to the French colony of Algeria in the early twentieth century.

14. On republican acknowledgment of Condorcet and other *philosophes*, see Miriam R. Levin, *Republican Art and Ideology in Late Nineteenth Century France* (Ann Arbor: UMI Research Press, 1986), 64, 66, 105, 238 n. 123; 250 n. 98; and Theodore Zeldin, *France, 1848–1945*, 2 vols. (Oxford: Oxford University Press, 1973–1977). Zeldin (1: 151) refers to the impact Condorcet's ideas had on the republicans, although Zeldin stresses their utopian character. The republicans acknowledged their intellectual debt to the *philosophes*: Lockroy, *La question sociale: Réponse à M. de Mun* (Paris: Balitout, Questroy et Ciel, 1883), 5; *Discours et opinions de Jules Ferry*, ed. Paul Robiquet, 7 vols. (Paris: A. Colin, 1893–1898), 6: 231–233; Simon, *Une académie sous le Directoire* (Paris: Calmann-Lévy, 1885).

15. See Landau, "La fabrication des rues de Paris," n.p.; Pinkney, *Napoleon III and the Rebuilding of Paris*, 72, 92; Michel Carmona, *Haussmann: His Life and Times, and the Making of Modern Paris* (Chicago: I. R. Dee, 2002). On his own efforts, see Georges Eugène Haussmann, *Atlas administratif des 20 arrondissements de la Ville de Paris* (Paris, 1868); *Mémoires du Baron Haussmann*; Plessis, *Rise and Fall of the Second Empire*, 63, 120.

16. Plessis, *Rise and Fall of the Second Empire*, 121–122. For a contemporary opposition point of view, see the scathing critique by Jules Ferry, *Contes fantastiques d'Haussmann: Lettre adressée à MM. les membres de la Commission du corps législatif chargés d'examiner le nouveau projet d'emprunt de la Ville de Paris* (Paris: A. Le Chevalier, 1868).

17. See Mayeur and Ribérieux, *The Third Republic from Its Origins to the Great War*, chapter 1, page 50, and chapter 2.

18. Jules Ferry, "Interpellation Langlois," 1884, quoted in Levin, *Republican Art and Ideology*, 1.

19. Quoted in Jones, *Paris: Biography of a City*, 332.

20. See Landau, sections "1870–1897, le temps de l'unification," and "Des ingénieurs liges de leur corps plus que des élus," in his article "La fabrication des rues de Paris," n.p. The article traces the growth and reorganizations of state and municipal services in charge of Paris, as well as of urban projects including schools, over the course of the various regimes of the nineteenth century. Also see Topalov, "Les 'réformateurs' et leurs réseaux," 11–58.

21. On scientists and science in the Third Republic, see for example Robert Fox and Anna Guagnini, introduction, and Fox, "France in Perspective," in Fox and Guagnini, eds., *Education, Technology, and Industrial Performance in Europe*, 1–10 and 201–226, respectively; Charles R. Day, *Education for the Industrial World: The Écoles d'Arts et Métiers and the Rise of French Industrial Engineering* (Cambridge, MA: MIT Press, 1987).

22. See Landau, "La fabrication des rues de Paris," for discussion of the negotiations between engineers and the Municipal Council for control over decision making. See also Nagaï, *Conseilleurs municipaux de Paris sous le IIIe République*; and Patricia R. Turner's review of Nagaï in *H-France Review* 4 (May 2004), no. 54, http://www.h-france.net/vol4reviews/turner.html, which summarizes his thesis as follows: "Although, as Nagaï repeatedly emphasizes, the Paris municipal council had little real power during the Third Republic, it nonetheless represented 'the place where diverse political and social interests were expressed and came into confrontation with one another.' During this period, Nagaï argues, the council reflected the intensification of political and electoral conflicts and continued opposition between Paris and the state." An alternative interpretation, Jones's *Paris: Biography of a City*, 331, 332–333, argues for continuity in the development of Parisian rebuilding under the Third Republic, proposing that, despite the reinstitution of the Municipal Council as an elected body after the Commune, the council remained under the tutelage of the prefect, who appointed the mayors of the *arrondissements*. However, bearing out Landau, the story of the Métropolitain is an example of engineers' continuing loyalty to their professional culture while working within the council's politically motivated desire to keep the railroads out of the city, as discussed later in the text.

23. Donatella Calabi, "Marcel Poëte: Pioneer of 'l'urbanisme' and Defender of 'l'histoire des villes,'" *Planning Perspectives* 11 (1996): 413–436.

24. On republican ideology, see n. 14 above and in addition: Jules Simon, *Dieu, patrie, liberté* (Paris: Calmann-Lévy, 1883); Édouard Lockroy, preface to Émile Monod, *L'Exposition universelle de 1889* (Paris, 1890), vol. 1, i–xxx. Useful secondary works include: Louis Legrand, *L'influence du positivisme dans l'oeuvre scolaire de Jules Ferry: Les origines de la laïcité* (Paris: Librairie Marcel Rivière, 1961); Roger Saltau, "Book 3: The Republican Era (1875–1914)," in *French Political Thought in the Nineteenth Century* (London: E. Benn, 1931); Raymond Delatouche, "Vers un renouveau de la physiocratie," *Journal de la Société de Statistique de Paris* 117 (1976): 47–53; Levin, *Republican Art and Ideology*, 147–175.

25. See note 16 above and Rondo E. Cameron, "The Crédit Mobilier and the Economic Development of Europe," *Journal of Political Economy* 61, no. 6 (December 1953): 461–488, http://www.jstor.org/stable/1825407.

26. Eva Etzioni-Halevy, *Bureaucracy and Democracy: A Political Dilemma* (London: Routledge, 1985), 32.

27. Levin, *Republican Art and Ideology*, 7–8.

28. See note 24 above.

29. Ibid.

30. See notes 14 and 24 above. Also David Thomson, *Democracy in France: The Third Republic* (London: Oxford University Press, 1946), 116–134, is especially good on the national vision.

31. See Levin, *Republican Art and Ideology*, 9–32.

32. Biographies of Jules Ferry include Maurice Reclus, *Jules Ferry: 1832–1893* (Paris: Flammarion, 1947); Jean Dietz, "Jules Ferry et les traditions républicaines," *Revue politique et parlementaire* 160 (July 1934). On Jules Simon, Levin's *Republican Art and Ideology*, 4, provides a short sketch; see also Léon Seché, *Jules Simon, sa vie et son oeuvre* (Paris: A. Dupret, 1887). On Ferry, Simon, and Lockroy, also see Levin, *Republican Art and Ideology*, 223–224, note 1.

33. The Loi Siegfried of 30 November 1894 gave working-class housing a special status under the title of "habitations à bon marché" (HBM). It relied on tax exemptions and other fiscal incentives to stimulate low-cost housing construction. The law of 14 December 1914 modified that of 29 May 1906. Together their objective was to further the aims of the 1894 law. Susanna Magri, *Les laboratoires de la réforme de l'habitation populaire en France: De la Société française des habitations à bon marché à la section d'hygiène urbaine et rurale du Musée social, 1889–1909* (Paris: Ministère de l'Équipment, du Logement, des Transports et du Tourisme, 1996).

34. On Poubelle, see Donald Reid, *Paris Sewers and Sewermen: Realities and Representation* (Cambridge, MA: Harvard University Press, 1991).

35. See Daly, "Toast aux ingénieurs," 101.

36. Landau, "La fabrication des rues de Paris," passim. Alphand published volumes rich with information about his work on embellishing Paris and on the 1889 exposition: A. Alphand, *Les promenades de Paris: Histoire—description des embellissements—dépenses de création et d'entretien des Bois de Boulogne et de Vincennes, Champs-Élysées—parcs—squares—boulevards—places plantées, étude sur l'art des jardins et arboretum* (1873; repr., Princeton: Princeton Architectural Press, 1984); and Adolphe Alphand, *Exposition universelle internationale de 1889. Direction générale des travaux. Rapport général sur les travaux de l'année 1887–1888* (Paris, 1889).

37. For biographical information on Picard, see "Alfred Picard," *Génie civil* 62, 33e année (1912–1913) and http://www.annales.org/archives/x/picard.html. Picard was a student at the EPT before going on to graduate from the EPC. He helped

direct telegraph, canal, and railroad projects during the Second Empire. Under the Third Republic, he held high posts in the Ministry of Public Works, and was a member of the Académie des Sciences. Among Picard's published works are *Exposition universelle internationale de 1889 à Paris: Rapport général par M. Alfred Picard*, 10 vols. (Paris: Imprimerie nationale, 1891–1896); *Les chemins de fer français: Étude historique sur la constitution et le régime du réseau; débats parlementaires, actes législatifs, réglementaires, administratifs, etc.* (Paris: J. Rothschild, 1884–1885); *Exposition universelle internationale de 1900 à Paris. Rapport général administratif et technique*, 8 vols. (Paris: Imprimerie nationale, 1902); *Le bilan d'un siècle (1801–1900)*, 6 vols. (Paris: H. Le Soudier, 1906–1907); and *Les chemins de fer: Aperçu historique* (Paris: Dunod et Pinat, 1918).

Frank Dobbin, "How Institutions Create Ideas: Railroad Finance and the Construction of Public and Private in France and the United States," *Theory and Research in Comparative Social Analysis* (29 January 2004), paper 11, http://repositories.cdlib.org/uclasoc/trcsa/11, cites Picard in describing how "the French gave credit to state finance and initiative and took the lesson that the state's role was to develop and capitalize important industrial projects, orchestrating the activities of private parties toward national goals (*Les chemins de fer: Aperçu historique*). These lessons would affect how each nation developed future industries" (1). Also see his citations on Picard on 6, 7, 15, 33, 36.

38. A relatively recent biography of Bienvenüe is Claude Berton and Alexandre Ossadzow, *Fulgence Bienvenüe et la construction du Métropolitain de Paris* (Paris: Presses de l'École Nationale des Ponts et Chaussées, 1998). A brief history of the Métro can be found in Norma Evenson, "The Paris Metro," in David C. Goodman, ed., *The European Cities and Technology Reader: Industrial to Post-Industrial City* (London: Routledge in association with the Open University, 1999).

39. One of the few biographies of this master entrepreneur-engineer is Henri Loyrette, *Gustave Eiffel* (New York: Rizzoli, 1985).

40. Charles Garnier, *L'habitation humaine* (Paris: Hachette, 1892); Frantz Jourdain and Charles Garnier, *Exposition Universelle de 1889: Constructions élevées au Champ de Mars par M. Ch. Garnier ... pour servir à l'histoire de l'habitation humaine; texte explicatif et descriptif* (Paris: Librairie Centrale des Beaux-Arts, 1892).

41. Directors of the Muséum d'Histoire Naturelle were Michel-Eugène Chevreul, 1863–1879; Edmond Frémy, 1879–1891; Alphonse Milne-Edwards, 1891–1900. Listed in R. Verneau, "Le professeur Hamy et ses prédécesseurs au Jardin des Plantes," *Anthropologie* 21 (1910): 257–279. Also helpful in providing historical context are Camille Limoges, "The Development of the Muséum d'Histoire Naturelle of Paris, c. 1800–1914," in Robert Fox and George Weisz, eds., *The Organization of Science and Technology in France (1808–1914)* (Cambridge: Cambridge University Press, 1980), 211–240; Alice L. Conklin, "Civil Society, Science, and Empire in Late Republican France: The Foundation of Paris's Museum of Man,"

Osiris, 2nd ser., 17 (2002): 263–266, and notes 3, 5, and 6; and J. Clifford, "On Ethnographic Surrealism," *Comparative Studies in Society and History* 23, no. 4 (October 1981): 539–564.

For a contemporary assessment of anthropology in Paris at the time of the 1889 exposition, see Otis T. Mason, *American Anthropologist* 3, no. 1 (January 1890): 27–36, who lists Hamy on the local arrangements committee.

42. Ernest-Théodore Hamy, *Les origines du Musée d'ethnographie* (1890; facsimile repr., Paris: J.-M. Place, 1988); Nelia Dias, *Le Musée d'ethnographie du Trocadéro: 1878–1908: Anthropologie et muséologie en France* (Paris: Éditions du Centre National de la Recherche Scientifique, 1991); Clifford, "On Ethnographic Surrealism."

43. There is a growing bibliography on the Musée Social. In addition to the Musée's own publications that appeared between 1894 and 1914 recounting its founding, goals, and activities, recent publications include: Horne, *A Social Laboratory for Modern France*; Ann-Louise Shapiro, "Housing Reform in Paris: Social Space and Social Control," *French Historical Studies* 12, no. 4 (Autumn 1982): 486–507, http://www.jstor.org/stable/286422; Topalov, "Les 'réformateurs' et leurs réseaux"; Sanford Elwitt, "Social Reform and Social Order in Late Nineteenth-Century France: The Musée Social and Its Friends," *French Historical Studies* 11, no. 3 (Spring 1980): 431–451, http://www.jstor.org/stable/286396.

44. Elwitt, "Social Reform and Social Order." On Hénard as an urbanist, see Peter M. Wolf, *Eugène Hénard and the Beginning of Urbanism in Paris, 1900–1914* (The Hague: International Federation for Housing and Planning; Paris: Centre de Recherche d'Urbanisme, 1968); Peter Wolf, "The First Modern Urbanist," *Architectural Forum* 127, no. 3 (October 1967): 52.

45. In addition to editing the *Album de Statistique Graphique* for over twenty years, Émile Cheysson authored numerous statistical studies on urban and national life, including *La circulation sur les routes nationales d'après les comptages de 1882* (1884); *La statistique géometrique: Ses applications industrielles et commerciales* (1887); *Statistique géometrique: Méthode pour la solution des problèmes commerciaux et industriels* (1887); and *L'affaiblissement de la natalité française* (1891). He also wrote a study of his mentor, *Frédéric Le Play, sa méthode, sa doctrine, son école* (1905).

46. Marcel Poëte, *Une vie de cité: Paris de sa naissance à nos jours*, 3 vols. and *album* (Paris: Auguste Picard, 1924–1931); and the better-known *Introduction à l'urbanisme: L'évolution des villes, la leçon de l'histoire, l'antiquité; ouvrage illustré de trente-deux phototypies*, 2nd ed. (Paris: Editions Anthropos, 1967).

47. Calabi, "Marcel Poëte: Pioneer of 'l'urbanisme,'" discusses the development of Poëte's ideas, pointing out that he produced extensive in-house memos, reports, and projects as well as published works from 1903 on through the books of the 1920s and 1930s for which he is best known today. This article encapsulates the

ideas she develops more fully in Donatella Calabi, *Marcel Poëte et le Paris des années vingt: Aux origines de l'histoire des villes* (Paris: L'Harmattan, 1998).

48. Calabi, *Marcel Poëte,* 420.

49. Ibid. See also references on 434–436, notes 19, 30–39. Poëte's prewar published works include *Formation et évolution de Paris* (Paris: F. Jouven, [1910]) and *La proménade à Paris au XVIIe siècle: L'art de se promener, les lieux de promenade dans la ville et aux environs* (Paris: Armand Colin, 1913). Beginning in 1904, Poëte taught a course at the Bibliothèque on the history of the city of Paris. Diana Periton, "Generative History: Marcel Poëte and the City as Urban Organism," *Journal of Architecture* 11, no. 4 (2006): 425–439, maps out his efforts to transform a library, the Bibliothèque Historique de la Ville de Paris, from a passive storage for historical documents to an active urban organism.

50. *Transactions* (London: Royal Institute of British Architects, 1911), 345–367. On Hénard's activities as an urbanist, see Eugène Hénard and D. A. Agache, *Création officielle de la SFAU* (1914); and Wolf, "The First Modern Urbanist," 52. According to Hénard and Agache, *Création officielle de la SFAU*, the Société Française des Urbanistes was founded in 1911 by D. A. Agache, M. Auburtin, A. Bérard, E. Hébrard, L. Jaussely, A. Parenty, H. Prost (architects), J. C. N. Forestier (engineer and landscape painter), and E. Redont (landscapist). These men were members of the Musée Social. In 1914 they founded the Société Française des Architectes Urbanistes. The SFAU's goal was to "réunir une documentation technique, de nouer et d'entretenir des relations avec les groupements similaires à l'étranger, de centraliser les voeux émis dans les derniers congrès urbains et d'en étudier la réalisation pratique, de participer aux expositions qui auront lieu tant en France qu'à l'étranger, de se tenir à disposition des intéressés pour toute consultation."

51. Between 1887 and 1914 Hénard published a number of studies on urban planning and exposition plans. Among those on the expositions are Eugène Hénard, *Étude sur une application du transport de la force par l'électricité: Projet de train continu (système breveté S. G. D. G.) pour l'Exposition universelle de 1889, destiné à obtenir la suppression de la fatigue des visiteurs* (Paris: Baudry, 1887); and *L'Exposition universelle de 1900 devant le Parlement, pourquoi il est nécessaire d'exécuter le projet issu du concours public de 1894 et des travaux du jury, par Eugène Hénard* (Paris: G. Delarue, 1896).

On his urban planning publications, see "The Cities of the Future," in Hénard, *L'Exposition universelle de 1900*; *E. Hénard: Rapports à la Commission des perspectives monumentales de la Ville de Paris* (Paris, 1911); and the series of short studies published in *Études sur les transformations de Paris* (Paris: Librairies-Imprimeries Réunis, 1903–1909).

52. For Hénard's solution to growing congestion on the streets—the invention of traffic circle in 1903—see Wolf, *Eugène Hénard*. For his drawing of an imagined

future urban system for Paris, see Hénard, "The Cities of the Future," fig. 4. More in the realm of projected planning based on opportunity for development is Jules Siegfried and Eugène Hénard, *Les espaces libres à Paris: Les fortifications remplacées par une ceinture de parcs. Jules Siegfried: exposé, projet de loi. Eugène Hénard: rapport technique* (Paris: A. Rousseau, 1909).

53. Among the vast number of studies on Paris during the Belle Époque, two of those most relevant to this chapter are Evenson, *Paris: A Century of Change*; and Charles Rearick, *Pleasures of the Belle Époque: Entertainment and Festivity in Turn-of-the-Century France* (New Haven: Yale University Press, 1986).

54. For Paris street construction statistics for the 1880s as well as the span of the nineteenth century, see André Guillerme and Sabine Barles, *Histoire, statuts et administration de la voirie urbaine: Guide pratique de la voirie urbaine* (Paris, 1999–2003), fasc. no. 1 (October 1998).

55. Landau, section "La voie publique, territoire de recherche et d'innovations," in "La fabrication des rues de Paris," n.p.

56. Jones, *Paris: Biography of a City*, 332.

57. See note 36 above, especially Alphand, *Les promenades de Paris* and *Exposition universelle internationale de 1889*.

58. The drawings and text describing the plans can be found in Alfred Picard, *L'art des jardins. Parcs—jardins—promenades—étude historique—principles de la composition des jardins—plantations—décoration pittoresque et artistique des parcs et jardins publics; traité pratique et didactique* (Paris: J. Rothschild, 1886), 329–340.

59. Landau, "La fabrication des rues de Paris"; André Guillerme, *Les corps sur la route: Les routes, les chemins et l'organisation des services au XIXème siècle* (Paris: Presses de l'École Nationale des Ponts et Chaussées, 1984); Reid, *Paris Sewers and Sewermen*, 62–63 and notes 50–69.

60. Reid, *Paris Sewers and Sewermen*, 62.

61. The following provide a picture of this growing urban technological net: Guillerme and Barles, *Histoire, statuts et administration de la voirie urbaine*; also André Guillerme, "Review of Helene Harter, *Les ingénieurs des travaux publics et la transformation des métropoles américaines, 1870–1910*, H-Urban, H-Net Reviews, August 2002. http://www.h-net.msu.edu/reviews/showrev.cgi?path=97901039514246. See also Joel A. Tarr and Gabriel Dupuy, eds., *Technology and the Rise of the Networked City in Europe and America* (Philadelphia: Temple University Press, 1988).

62. Guillerme, *Les corps sur la route*, 24.

63. Histories of these various institutions and their growth (or lack thereof) during this period: Fox, "France in Perspective," provides an overview; André Grelon,

"The Training of Engineers in France, 1880–1939," in Claudine Fontanon and André Grelon, eds., *Les professeurs du Conservatoire national des arts et métiers: Dictionnaire biographique, 1794–1955* (Paris: Institut National de Recherche Pédagogique, 1994), 43–46; Karine Dubreuil and Sébastien Beltramini, *Le Conservatoire national des arts et métiers: Du Prieuré Saint-Martin au Conservatoire,* Mémoire rédigé sous la direction de Mr. Jean Castex (Paris: École d'Architecture de Versailles, 1999), 36 ff; Day, *Education for the Industrial World,* 20–21, 235–237; Girolamo Ramunni, *1894–1994: Cent ans d'histoire de l'École Supérieure d'Électricité* (Paris: École Supérieure d'Électricité, 1994).

There are several volumes on the history of the university and the Sorbonne, including the division (curricular, spatial, professorial, and architectural) between literature and science that developed over this period: André Tuilier, *Histoire de l'Université de Paris et de la Sorbonne* (Paris: Nouvelle Librairie de France, 1997), vol. 2; Jean-Louis Leutrat, *De l'Université aux Universités* (Paris: Association des Universités de Paris, 1997); Philippe Rivé, *La Sorbonne et sa reconstruction* (Lyon: La Manufacture, 1987). These divisions at the university are summed up on the official web page of the Sorbonne: "Nénot [the architect], en revanche, sait trouver dans ses travaux d'école les éléments qui assureront la cohérence et la (relative) lisibilité du nouvel ensemble, ordonné par une systématique dualité de ses éléments: aux deux principales facultés hébergées, les lettres et les sciences, correspondent deux espaces distincts, de part et d'autre de la chapelle; les galeries, les axes, les amphithéâtres, les tours (physique et astronomie), jusqu'au décor et aux statues (Hugo et Pasteur …), tout va par deux et évoque alternativement les lettres et les sciences. Ces deux pôles sont en outre distingués par les matériaux: au grand appareil de pierre de taille qui règne de la rue des Écoles à la galerie Gerson répondent la brique émaillée et l'affichage d'un rationalisme plus moderniste pour la faculté des sciences."

64. For the institutional history of CNAM, see Day, *Education for the Industrial World*; and Fontanon and Grelon, introduction to *Les professeurs du Conservatoire national des arts et métiers.*

65. For a history of Supélec, see Ramunni, *1894–1994.*

66. The history of Thomson-Houston in France has yet to be written. Three sources, which I have not been able to consult, may be useful: "Procès verbaux du Conseil d'Administration," Thomson-France, Archives de Thomson (239, bd. Anatole-France, St.-Denis); Jacques Mars, "Essai de monographie de la Compagnie Française Thomson-Houston" (1937); and *Historique Thomson: Le Groupe de 1893 à 1977,* internal history (Paris: Thomson, 1979, not published), vol. 1.

67. Both the Métro and the Haussmann boulevards were finally completed in the 1920s and 1930s. As for modern housing for the workers, their construction had to wait until after World War II. On the history of the HBM, see Elwitt, "Social Reform and Social Order"; Lucien Lambeau, *Ville de Paris. Monographies. Les loge-*

ments à bon marché. Recueil annoté des discussions, délibérations et rapports du conseil municipal de Paris (Paris, 1897); Evenson, *Paris: A Century of Change*, 203–205; Roger H. Guerrand, *Les origines du logement social en France* (Paris: Éditions Ouvrières, 1966), 202–221; Ann-Louise Shapiro, *Housing the Poor of Paris* (Madison: University of Wisconsin Press, 1985), 56, 75–76; Octave DuMesnil and Charles Mangenot, *Enquête sur les logements, professions, salaires, et budgets* (Paris, 1899). The following has a good introductory chapter on changes in Parisian transportation, housing conditions, and social geography: Tyler Stovall, *The Rise of the Paris Red Belt* (Berkeley: University of California Press, 1990), http://ark.cdlib.org/ark:/13030/ft5r29n9vt/, 20–25.

68. The prefaces and introductions to the official reports for each exposition are good resources for statements on the innovative nature of organizers' intentions and the organizational and symbolic significance they invested in the expositions. On arrangements for the 1889 exposition, see Levin, *Republican Art and Ideology*, 61–62.

69. See Picard, *Exposition universelle internationale de 1889*; Picard, *Le bilan d'un siècle*; Lockroy, preface to Monod, *L'Exposition universelle de 1889*.

70. *Harper's Guide to Paris and the Exposition of 1900* (New York: Harper & Brothers, 1900).

71. Reid, *Paris Sewers and Sewermen*; Guillerme, *Voirie urbaine*, n.p.

72. For more on tramway and rail connections for Orsay and other rail stations in Paris during the exposition, see Picard, *Exposition universelle internationale de 1900*, 8: 401, 405. For similar arrangements during the 1889 exposition, see Picard, *Exposition universelle internationale de 1889*, 3: 261–262.

73. A useful modern history of the Paris Métropolitain is Roger-Henri Guérand, *L'aventure du Métropolitain* (Paris: Découverte, 1986). A number of other recent studies on the Paris Métro have appeared in French, among them A. Bindi and D. Lefeuvre, *Le Métro de Paris: Histoire d'hier à demain* (Rennes: Ouest-France, 1990); François Gasnault and Henri Zuber, *Métro-Cité: Le chemin de fer métropolitain à la conquête de Paris, 1871–1945* (Paris: Musées de la Ville de Paris, 1997); Gaston Jacobs, *Le Métro de Paris: Un siècle de matériel roulant* (Paris: Éditions La Vie du Rail, 2001).

74. Gabriel P. Weisberg, "The Parisian Situation: Hector Guimard and the Emergence of Art Nouveau," in Paul Greenhalgh, ed., *Art Nouveau: 1890–1914* (New York: Harry N. Abrams, 2000), 264–273.

75. On Picard's plan to integrate the Left and Right Banks via a bridge at the Esplanade des Invalides at the time of the exposition, see Alfred Picard, "Programme du concours sur les dispositions générales des bâtiments, jardins et agencements de l'Exposition (partie urbaine)," article 4, in Picard, *Exposition universelle internationale de 1900 à Paris. Rapport général administratif et technique. Pièces annexes.*

Actes officiels. Tableaux statistiques et financiers (Paris: Imprimerie Nationale, 1902–1903), 79, article 7.

76. For Hénard's contribution, see Hénard, *L'Exposition universelle de 1900*; Wolf, *Eugène Hénard*, 29; and Wolf, "The First Modern Urbanist," 52.

77. Debora L. Silverman, *Art Nouveau in Fin-de-Siècle France: Politics, Psychology, and Style*, Studies on the History of Society and Culture, 7 (Berkeley: University of California Press, 1989), 169. Evenson, *Paris: A Century of Change*, 24–34, 66–68, and figure 16, discusses Hénard's plans and describes his traffic circle idea. Hénard's later traffic circle plans include that for the Étoile constructed in 1907. This "carrefour à gyration" was an ingenious attempt to deal with the existing congestion of carriages, horses, and with the automobiles he predicted would soon increase the snarl.

78. Lockroy, preface to Monod, *L'Exposition universelle de 1889*, xv and xx; and Levin, *Republican Art and Ideology*, 56 and 249, note 94.

79. See for example, Picard, *Exposition universelle internationale de 1900. Pièces annexes*, arts. 9–13, 84–93, and tableaux 7–12, 739–910.

80. Eiffel also received some attention for his role in another project: the iron interior of the Statue of Liberty, whose head was on view in the Trocadéro Garden.

81. Gustave Eiffel, *La tour de trois cent mètres* (Paris: Lemercier, 1900).

82. The Eiffel Tower "résumait la grandeur et la puissance industrielle du temps présent. Sa flèche immense, en s'enfonçant dans les nuages, avait quelque chose de symbolique; elle paraissait l'image du progrès tel que nous le concevons aujourd'hui: spirale démesurée où l'humanité gravite dans cette ascension éternelle." Lockroy, preface to Monod, *L'Exposition universelle de 1889*, xxv (quoted in Levin, *Republican Art and Ideology*, 45).

83. The details of the contract are discussed in Picard, *Exposition universelle internationale de 1889*, 2: 266–269; *Harper's Guide to Paris*, 150–151; and Levin, *Republican Art and Ideology*, 42–43 and 239, note 130.

84. See Levin, *Republican Art and Ideology*.

85. Lockroy, preface to Monod, *L'Exposition universelle de 1889*, xxv; quoted in Miriam Levin, *When the Eiffel Tower Was New: French Visions of Progress at the Centennial of the Revolution* (South Hadley, MA: Mount Holyoke College Art Museum, 1989), 22, note 34. Electronic reproduction: Boulder, Colorado: NetLibrary, 2000. Available via the World Wide Web.

86. Picard, *Exposition universelle internationale de 1900*, 1: 10–12. Picard elaborated on this point in a later publication: his introduction to *Le bilan d'un siècle*, 1: i–iv.

87. Silverman, *Art Nouveau*, 160, 169, 284–289.

88. *Harper's Guide to Paris*, 164.

89. Lockroy, cited in Levin, *Republican Art and Ideology*, 108 and 251, note 117.

90. *Harper's Guide to Paris*, 156 ff.; Picard, *Exposition universelle internationale de 1900*; Picard, introduction to *Le bilan d'un siècle*.

91. Picard, *Le bilan d'un siècle*.

92. See explanations and lists of exhibitors in *Groupe XVII Colonisation, Classes 113 à 115, Catalogue Général Officiel. Exposition internationale universelle de 1900* (Paris: Imprimeries Lemercier, 1900), vol. 19; M. J. Charles-Roux, *Rapport Général. L'organisation et le fonctionnement de l'exposition des Colonies et Pays de Protectorat, Exposition universelle de 1900* (Paris: Imprimerie Nationale, 1902); *Guide illustré de l'Exposition coloniale française au Trocadéro en 1900* (Cambrai: F. et P. Deligne, 1900); and *Harper's Guide to Paris*, 167–168. On colonial exhibits at the 1889 exposition, see Louis Fontaine, *Rapports de M. Louis Fontaine, Ministère du Commerce, de l'Industrie et des Colonies; Exposition universelle internationale de 1889, à Paris* (Paris, 1891).

93. See Picard, introduction to *Bilan d'un siècle*; and Michael A. Osborne, "The Société Zoologique d'Acclimatation and the New French Empire: Science and Political Economy," in Patrick Petitjean et al., eds., *Science and Empires: Historical Studies about Scientific Development and European Expansion* (Dordrecht: Kluwer Academic Publishers, 1992), 299–306, which raises the question of how much credit to give to Paris as the inaugural point for such circulation.

94. Foreign Correspondent, "System for Death Statistics: International Commission Wants Adoption of Bertillon Method," *New York Times*, 2 September 1900.

95. Quote taken from A. Lawrence Rotch, "The International Congresses of Meteorology and Aeronautics at Paris," *Science* 12, no. 308 (1900): 796–799.

96. On Paris during World War I, see J. M. Winter, *Capital Cities at War: Paris, London, Berlin, 1914–1919* (Cambridge: Cambridge University Press, 1997).

97. For information on and descriptions of the social economy exhibits at the 1900 exposition, see for example *L'Économie sociale à l'exposition universelle internationale de 1900. Livre d'or des exposants du groupe XVI, Exposition universelle internationale de 1900 Jury international* (Paris: A. Rousseau, 1903), which lists the prizes and winners. Shawn Michelle Smith, introduction to *Photography on the Color Line: W. E. B. Du Bois, Race, and Visual Culture* (Durham: Duke University Press, 2004), 168, note 39, quoting *Harper's Guide to Paris*, 163, writes: "According to Harper's Guide to Paris, the Palace of Social Economy resembled 'an elegant and distinguished palace of the eighteenth century.' The structure was built entirely by workmen's associations. 'The dimensions are 328 feet by 110 feet; the height being 69 feet above the level

of the quay and 85 feet above the level of the Seine. The Palace is divided into two parts; on the ground floor are rooms in which the social economy exhibits are placed; the first floor is entirely reserved for the congresses. The building contains an enormous hall, 328 feet long and 39 feet wide, which precedes the meeting rooms, and which has its entire façade on the Seine. There are five halls for the congresses, one holding 800 persons, two smaller ones accommodating 250 each, and two others having a seating capacity of 150 persons.'" For social economy at earlier expositions during this period, set in a discussion of the broader political context, see Shapiro, "Housing Reform in Paris," 489, 493–494, 497.

98. See Horne, *A Social Laboratory for Modern France*; Topalov, "Les 'réformateurs' et leurs réseaux," 11–58; Rabinow, *French Modern*, 95, 182–183.

99. See the following essays in Colette Chambelland, ed., *Le Musée social en son temps* (Paris: Presses de l'École normale supérieure, 1998): Colette Chambelland, foreword; Pierre Rosanvallon, "Préface: Figures et méthodes du changement social," 7–8, and "Des réseaux et des champs d'intervention du Musée social"; Roger-Henri Guerrand, "Jules Siegfried, La Société française des habitations à bon marché et la loi du 30 novembre 1894"; Susanna Magri, "Du logement monofamilial à la cité-jardin: Les agents de la transformation du projet réformateur sur l'habitat populaire en France 1900–1909."

100. See notes 44, 50–52, 76, and 77 above on Hénard.

101. Rearick, *Pleasures of the Belle Époque*.

102. See description in *Harper's Guide to Paris*, 161–162.

103. Ibid., 162.

104. John Pickstone, "Museological Science? The Place of the Analytical/Comparative in Nineteenth Century Science, Technology and Medicine," *History of Science* 32 (1994): 111; also 118–124.

105. Lockroy, preface to Monod, *L'Exposition universelle de 1889*, contains a reprint of the artists' petition as well as the text of Lockroy's cuttingly droll reply.

106. Miriam R. Levin, "The City as a Museum of Technology," *History and Technology* 10, nos. 1–2 (1993): 27–36.

107. On the history of the Musée des Arts Décoratifs and its origins, see Yvonne Brunhammer, *Le beau dans l'utile: Un musée pour les arts décoratifs* (Paris: Gallimard, 1992). Hamy, *Les origines du Musée d'ethnographie*, 303–306, provides an idea of Jules Ferry's and Antonin Proust's efforts in the Ministry of Education and Fine Arts to found museums during this period.

108. *Catalogue des collections du Conservatoire national des arts et métiers, premier fascicule*, 8th ed. (Paris, 1905).

109. Dias, *Le Musée d'ethnographie du Trocadéro*; Hamy, *Les origines du Musée d'ethnographie*. To give a sense of the popularity of the type, Dias (p. 94) states that 78 ethnographic museums were founded in France between 1841 and 1860, and 92 between 1861 and 1880.

110. See Hamy, *Les origines du Musée d'ethnographie*, 303–306 for documents and correspondence regarding Ferry.

111. For an account of these institutions and Ferry and Proust's involvement in the society and the museum, see Brunhammer, *Le beau dans l'utile*, 33–37, 45. The UCBAI and SMAD were fused into the Union Central des Arts Décoratifs in 1882, created by government decree as an *association reconnue d'utilité publique*.

112. Dias, *Le Musée d'ethnographie du Trocadéro*, ii.

113. For example, photographic exhibits of French mechanical engineering from the 1900 exposition and the Political Economy exhibit from the 1889 exposition went to CNAM: see *Catalogue des collections du Conservatoire national des Arts et Metiers*, "Avis" and 45–46.

114. Hamy, *Les origines du Musée d'ethnographie*.

115. Dias, *Le Musée d'ethnographie du Trocadéro*, 105–109.

116. This claim to co-opt and link the subject matter of the other human sciences was matched by Hamy's classification system, which attempted to bring all of human production and invention over time and space, including that of colonized people, into one coherent explanatory order. Hamy explained: "Ethnography was the study of all the material manifestations of human activity" and its focus was on "all that which, in the material existence of individuals, families, or societies, bears a characteristic trait [of humanity]." ("L'étude de toutes les manifestations matérielles de l'activité humaine … tout ce qui, dans l'existence matérielle des individus, des familles ou des sociétés, présente quelque trait bien caractéristique, est du domaine de l'ethnographie.") Hamy, "Classement général des groupes et projet de classification des collections etographique," quoted in Dias, *Le Musée d'ethnographie du Trocadéro,* 154–155.

117. See Dias, *Le Musée d'ethnographie du Trocadéro*, on Hamy's classification system.

3 FROM MODERN BABYLON TO WHITE CITY: SCIENCE, TECHNOLOGY, AND URBAN CHANGE IN LONDON, 1870–1914

SOPHIE FORGAN

The term "Modern Babylon" nicely sums up the ambivalence with which many Victorians regarded their capital city; its teeming multitudes, babel of languages, and great buildings evoked both the false gods and great empires of antiquity.[1] Historians have found it a fruitful analogy, and it serves as a convenient point of departure for this study of London. On the other hand, "White City," an exhibition site named in homage to the famous 1893 Chicago world's fair, embodied the rather more positive attitudes toward the city that had emerged by the latter part of the period under discussion in this chapter. London was an ancient city, but one that was in the throes of becoming a modern metropolis and increasingly conscious of its position as imperial capital. This threefold character underlies its development during this period of intense scientific and technological change.

London was huge, complex, and dynamic, geographically the largest of the cities under consideration in our study, and throughout the nineteenth and most of the twentieth century, the largest European city in terms of population as well.[2] Moreover, industrialization had taken place earlier in London—and generally in Britain—than in the other countries studied here, so that by 1840 it had become a center for skilled artisanal and small-scale industrial works.[3] Over the course of the nineteenth century, industry—armaments, shipbuilding, food processing, and chemicals—continued to expand with the growth of the city. In the early decades of the twentieth century, new industrial concerns, such as the automotive manufacturer Ford and producers of domestic appliances, established themselves in the suburbs, in areas which were geographically close and would soon be subsumed within Greater London.[4]

But while there was significant structural change, London experienced neither the abrupt economic ruptures that struck many Continental cities where industrialization occurred later, nor a natural disaster, such as the Chicago fire that wiped out much of that city. New industries and adaptability insulated London, though that is not to say that there were no periods of grave difficulty, for the city suffered the effects of economic cycles from the 1860s on. Industrial strife was certainly present too, for example the well-known match girls' strike, the 1889 dock strike, and a bitterly fought nationwide engineering strike in 1897; later, the strike-prone years immediately following World War I culminated in the General Strike of 1926. During the 1880s, a period of intense social concern, great enquiries into poverty were begun by social investigator Charles Booth, and the notion of "poverty" itself began to be redefined.[5] But compared to the Continent or to Chicago, London was relatively stable in terms of class and industrial conflict during the period from 1870 to 1930, when the conflicts that did occur were essentially nonrevolutionary.[6]

London was widely admired in the middle of the nineteenth century for its provision of public services, with French and German commentators praising its sewer system, water supply, and slum clearance. Albert Shaw, the American journalist and editor who in the 1890s critiqued American municipal government, spent considerable time examining London and its municipal government agenda, writing that the city "must convey some direct lessons, and must throw many side-lights upon the treatment of corresponding metropolitan problems everywhere."[7] In London, however, the crucial characteristic was that most municipal services developed in a disconnected, laissez-faire way, with numerous private companies competing to provide tramways and underground trains, water, gas, and electricity. Because of the city's size and the multiplicity of vested interests, London's municipal government was slower than the other cities in this book to take ultimate control of power, transport, and water.

During the last quarter of the nineteenth century, change and progress were taken largely for granted.[8] This chapter explores the ways that such ideologies were part of a larger culture of change, and how this culture transformed the city and attitudes toward it.[9] Like those of other cities, London's urban elites were intensely aware of comparable developments in other cities.[10]

This chapter is divided into four main sections. The first focuses on identifying and characterizing the relevant urban elites. This is problematic in London, due to the absence of general studies on its elites in this period; the size and complexity of the city meant that professional leaders were varied and relatively dispersed. A new categorization of elites is adopted here in order to illuminate the technological, social, and cultural context of London. The second section examines how these elites became embedded in the fabric of the city's life and culture. The third explores those institutions which embodied and disseminated science and technology for wider public consumption: museums that presented a guide to modern city living, and those collections that included both the glories of the past and the inventions of the present. Finally we focus on how the future was invoked at London exhibitions, where the latest technologies were pressed into service. Exhibitions were also key sites for displaying continuities with the past, portrayed in order to demonstrate progressive improvement and to provide picturesque delight. London, after all, as Dr. Johnson had said long before, "comprehend[ed] the whole of human life in all its variety, the contemplation of which is inexhaustible."[11] London's culture of change in this period was characterized by enthusiasm for technological modernity (accompanied by a sense of progressive evolutionary development), a devotion to free enterprise and respect for local autonomy, and a keen attachment to historic continuity.

PART I   ELITES IN LONDON: INDIVIDUALS, EXPERTISE, AND NETWORKS

Nothing would be more fatal than for the Government of States to get in the hands of experts. Expert knowledge is limited knowledge: and the unlimited ignorance of the plain man who knows where it hurts is a safer guide than any vigorous direction of a specialised character. Why should you assume that all except doctors, engineers, etc., are drones or worse?

—Winston Churchill to H. G. Wells (1902), quoted in Harold Perkin, *The Rise of Professional Society: England since 1880*

Historians divide British elites in the late Victorian period into three main types: the London-based commercial and financial elite, which became important in the later nineteenth century; the landed elite, many of whom during this period continued their seasonal migrations between the capital and the country; and the northern industrial elite, who are not a part of this account.[12] Few London scientists or technologists belonged to the commercial or landed elites, with the notable exceptions of Sir John Lubbock (1834–1913, polymath and financial expert, later Lord Avebury) and John Strutt, Lord Rayleigh (1842–1919, physicist and cousin of A. J. Balfour, who was Prime Minister from 1902 to 1905).[13] London elites were quite unlike the political, social, and industrial elites in other British cities, who as a group were "coherent, integrated and based in the locality" through family links, intermarriage, or common membership at church or chapel.[14] While one can certainly find family connections and common memberships among its elites, London was simply too big for the sort of coherent and integrated elite found, for example, in the Chicago Commercial Club.

The prime function of elites was—and is—to govern and to influence.[15] Because of London's size, governance operated at a number of levels. London was the imperial capital, the capital of the nation, *and* a major metropolis. It had a unique form of governance—notable, many critics would say, for its inability to grapple with the monstrous problems confronting the city, though this failure was mitigated somewhat following the formation of the London County Council in 1889, which created the main structure for modern city government.[16] Compared to Berlin, royal influence was relatively slight in London during the latter decades of the nineteenth century, though royals acted as patrons of museums, colleges, and new educational buildings, and royal parties regularly visited the exhibitions discussed later in this chapter. But royal patronage was not a driver of urban change as it had been before Prince Albert's death in 1861, when there had been royal involvement in the development of the South Kensington complex of museums and schools.

While there were overlaps, elites involved in imperial and national government tended to be separate from those concerned with municipal services such as street lighting, gasworks, sewers, and education. Scientific expertise applied to urban issues was the province of relatively lowly professionals such as municipal engineers, gas inspectors, and water analysts, although in the latter part of the century scientific experts became prominent as court witnesses and public activists.[17] That progression has generally been explained in terms of increasing specialization and professionalization. Those whom historians typically regard as belonging to

London's scientific elite at the start of our period in 1870 were Fellows of the Royal Society and the other major learned societies, the expanding scientific professoriat of the colleges of London University, and networking groups such as the well-known X Club.[18] By 1900, however, the question of who formed the capital's scientific elite had become more complex.

In this chapter we analyze elites from a different angle from that used elsewhere in this volume, picking out three types of scientists or technologists who can be defined by their roles in the corporate structures from which elite authority would increasingly derive. These types are, first, the highly individualistic scientist, characteristic of the earlier nineteenth century, who was not necessarily attached to one particular institution but might be active in several learned societies or similar bodies; second, the public servant, who worked in a local or state-funded institution; and third, the "intellectual entrepreneur," whose position in an organization enabled him to play an important role in metropolitan affairs. These divisions reflect the shift in social structures from a closely networked elite who functioned as "social leaders" in a locality to more modern "public persons" engaged in governance.

Those in the first category might be academic scientists, or independent gentlemen such as Rayleigh and Lubbock, whose reputations and interests were national rather than local. Sir John Lubbock, youngest member of the X Club, was the only one to play a direct role in London politics, serving on the London County Council briefly as its chairman, and also as a Member of Parliament, first for Maidstone and then, beginning in 1880, for the University of London.[19] Lubbock was first and foremost a banker who represented City of London financial elites, as well as being a politician and man of science. With his wealth and learning, Lubbock moved easily between public affairs, parliamentary duty, business and commerce, and the world of learned scientific societies, gathering "an alphabet of Honorary Degrees and memberships" as he went.[20] His frequent appearances upon the public stage were marked by tact, good manners, and a seemingly effortless grasp of his subject. He was highly respected as an expert mediator, and his political views remained broadly liberal.[21] Lubbock was certainly forward-looking in his support of modern technological private enterprise, underlined by his joining the board of the Edison Company in London in 1882, where he helped to negotiate the merger of Edison and Swan shortly thereafter.[22] During his most active years, before 1906, there was hardly a major initiative in which he was not in some way involved, if only to add his name to the list of supporters. Always welcome on any organizing committee, he was president at various times of no fewer than ten learned societies, as well as four international associations and three City of London commercial institutions. Since he was tutored in Darwinian science by the great man himself, it is not surprising that Lubbock's overall political views drew on notions of adaptive change as much as the idea of straightforward progress.[23]

The second type of elite London scientist was the public servant, who often had a military or an engineering background (and sometimes both), and owed his position in urban affairs to the corporate structure within which he functioned; such men were "civil scientists," members of a class that had started to emerge over the previous quarter century.[24] They had to abide by the rules governing employees, particularly in the civil service, but at the same time they aspired to professional status, which for some still implied notions of gentlemanly background and behavior.[25] As servants of the state, professionals of this type were bound up with national institutions, chiefly located in London: the Post Office, the Department of Science and Art and its museums and colleges, government departments such as the Board of Education or Office of Works, and later in the period the National Physical Laboratory in the suburbs at Teddington. Because of their particular expertise, they might also be active in new professional institutes such as the Institution of Electrical Engineers, and on retirement were often involved in companies providing municipal services. Railway and electrical supply companies were happy to invite such men to become directors.

A typical example was W. H. Preece (1834–1913), electrician and chief engineer to the Post Office, who was very active in the Institution of Electrical Engineers and was involved in many metropolitan organizations. His activities are discussed below in greater detail. Another prominent public servant, Sir John Donnelly (1834–1902), was Director of Science and then head of the Department of Science and Art, responsible for the South Kensington Museum and the scientific schools associated with it. He remained a key figure in the promotion of technical education and in London scientific life until his retirement in 1896. Douglas Galton (1822–1899), an engineer and sanitary expert, served in the War Office and then the Office of Works, where he was director of public works and building until retirement in 1875. A man of boundless energy, Galton was especially active in the Society of Arts, serving on its council and writing in the society's journal. He served on the City Livery Companies' committee on technical education in 1878 and participated in exhibitions relating to health. He sat on the Council and Hospital Committee of University College London and campaigned for the establishment of the National Physical Laboratory, which was set up in 1899. Richard Glazebrook (1854–1935), the National Physical Laboratory's first director, ensured that the laboratory became the center for research on electrical engineering, metrology, shipping, and aeronautics, as well as the establishment of standards across several scientific fields. Glazebrook took a prominent role in the Royal Society and in other London learned societies and professional institutions, was involved with the development of the Science Museum from 1910, and served as an 1851 Commissioner, responsible for awarding science scholarships among many other activities. These men—with the possible exception of Glazebrook—have not been regarded as doing scientific research of the first rank, but the state institutions to which they belonged provided a base from which they exercised a great deal of authority and influence.[26]

The third category of metropolitan elites includes men who were essentially intellectual entrepreneurs, located in institutions at the heart of metropolitan life, and passionately devoted to science and the promotion of scientific and technical education. One such man was Henry Trueman Wood (1845–1929), who served as secretary of the Society of Arts from 1879 to 1917. The Society of Arts is often omitted from histories of scientific culture in London, though it is featured in Bernard Becker's *Scientific London,* published in 1874, immediately following the entries on the Royal Society and Royal Institution, which gives some indication of its contemporary status.[27] The society was conveniently located in the Adelphi, a few minutes from Trafalgar Square and the clubs and institutions of the West End. Under the patronage of Prince Albert it had undergone a marked revival through its role in organizing the 1851 Great Exhibition, and it continued to be a center for discussions, lectures, and meetings, frequently lending its rooms to other bodies. It published a weekly journal, awarded medals and premiums for practical inventions, organized a series of technical qualifications, and lobbied vigorously for improved educational provision and on matters such as patent law. It had a large membership, including a significant number of residents abroad in parts of the British Empire, and maintained its interest in imperial affairs through its Indian Section. Moreover, the Society of Arts was extremely important in organizing exhibitions.[28]

Trueman Wood provides a key link between different groups of elites. He maintained a long association with the British Association for the Advancement of Science, was frequently called upon to assist in organizing exhibitions, and was very active in the promotion of technical education. A passion for photography connected him with a new industry, and he became a director of Kodak, Limited, as well as president of the Royal Photographic Society from 1884 to 1896. Among Wood's circle of friends were many who had a long-standing interest in the Society of Arts, served on its council, lobbied for its causes, used its lecture theater and journal to promote their views, and found its offices a congenial place to gather. He had links with the Department of Science and Art, through Donnelly and his successors, as well as with the Imperial Institute after it opened in 1893. Wood was extremely close to some of the City Livery Companies, which would provide much of the financing for technical education. He ensured that the council of the Society of Arts contained men such as Preece and Galton, mentioned above, as well as engineers and technologists who had important connections with other metropolitan organizations. These included another entrepreneurial type, Frederick Bramwell (1818–1903), a mechanical engineer and technical legal expert, who belonged to the Goldsmiths Company and represented it on the City and Guilds of London Institute for technical education, as well as serving as secretary of the Royal Institution from 1885 to 1900.

Trueman Wood seems to have been particularly successful in engaging some of the eminent engineers who were involved in metropolitan works or consultancy.[29] He had friends among journalists, wrote for the *Graphic* and the *Daily Graphic*, and was equally at home with his "cronies" at the Athenaeum.[30] While Trueman Wood was not the leader of any formal group, during his years as its secretary, the Society of Arts was concerned with activities that promoted science and technology, and generally made a significant contribution to the city's institutions and metropolitan life. Its place as an influential body may have been only temporary, but for a time the society was a key forum for the city's scientific and technological elites.

This taxonomy of London's elites is by no means complete, but serves to expand the concept of an elite composed principally of academic research scientists to encompass the range of public servants and civil scientists and their links across different organizations, fostered by men such as Trueman Wood. There were, of course, other such sites and organizations in London, and few were totally exclusive. Lyon Playfair, for example, formerly an academic chemist, became an MP in 1868, and was frequently chairman of commissions and committees, close to politicians and deeply involved with South Kensington institutions.[31] Norman Lockyer was another entrepreneurial figure. An astronomer and editor of the weekly journal *Nature*, he was secretary to the Devonshire Commission on scientific education (1870–1875); later he was employed at South Kensington in charge of the solar physics laboratory, and was involved in London scientific affairs, though frequently embroiled in controversy.[32]

London clubs and learned societies continued to provide a central nexus of communication and sociability, a tradition continued by newly founded professional institutions. Numerous professional groups pursued their own interests, whether chemistry, water analysis, architecture, or any of the numerous sub-branches of engineering. Although conflicts between interest groups occurred,[33] the sprawling metropolis provided both expansive opportunities and such intractable problems that individual disputes pale by comparison. We now need to examine how scientific elites both attempted to create and respond to change in the growing metropolis.

PART II   LONDON: DEFINING MODERNITY AND ENGAGING ELITES

I would remark that, topographically, Modern London is essentially Protean, and there can be no finality in its depiction. To "improve" is the order of the day. On every side, familiar buildings, favourite nooks and corners, and long-cherished haunts are being transformed out of all recognition.

—Arthur H. Beavan, *Imperial London* (1901)

Scientific elites took little part in the direct governance of the city, notwithstanding the case of Sir John Lubbock, who arguably was regarded as acting for the commercial elite of the City. In many respects scientific elites might be described as being *in* the city rather than *of* the city.[34] In part this was a result of the fact that elites were relatively dispersed in the vast metropolis. It was also a result of the laissez-faire development of London's infrastructure: private water, transport, and electricity companies competed alongside municipal providers of the same services in some areas. There was little technological systems-building, as was evident in the chaos of the provision of electricity throughout the metropolis. Moreover, there was hostility to the creation of integrated services if they were to be run by the borough councils and financed through the public purse. A number of elites, notably Lubbock, were opposed to public provision of services on the grounds that it would mean that the LCC engaged in what was termed "municipal trading," considered to be a monopolistic practice contrary to the principles of free trade—which raised the dread specter of socialism. "Municipal trading" was believed to threaten private property and investment made, for example, in the plant and machinery of power stations. Many in the business and engineering fraternity therefore had a vested interest in opposing this aspect of technological modernity, even as the city grew relentlessly in size.

Nevertheless, many in the scientific and technological elites defined earlier both benefited from the city's growth and helped to shape its development. For both private enterprise and municipal services companies provided a rich area for expert consultancy. Frederick Bramwell, mechanical engineer and friend of Trueman Wood, was retained by all eight metropolitan water companies, as well as supervising the construction of works for a West End electric lighting company.[35] Edward Frankland, chemist and colleague of T. H. Huxley at the South Kensington College of Science, was frequently consulted on water purity, clashing on occasion with other experts appointed by water companies.[36]

As the state agencies (in which civil scientists served) and the professional institutes (home to the engineers) expanded their physical presences within the city, the new professional elites sought to claim their place in metropolitan scientific life. One way of achieving this was to demonstrate modernity by emphasizing the forward march of technology and the importance of international links to the imperial capital; in addition, the elites took all available opportunities to provide educational and enlightening displays about their fields of expertise. At the same time, it was essential to provide a face for the new technologies that would be acceptable to people living in the rapidly changing city. These aspects will be examined through two key institutions—the Post Office and the Institution of Electrical Engineers—and through ideas about appropriate architecture for new urban technological facilities.

THE GENERAL POST OFFICE

The central General Post Office, housed in three huge adjoining buildings in the City, was naturally a center for the development of new technology, if only to maintain the empire's global communications. It was thus able to exert pressure on government for additional funds or space, despite the Treasury's parsimony. In 1885, for example, expansion of the Post Office was favored over extensions to the South Kensington Museum.[37] The Post Office also had the largest bank in the world, which necessitated its developing the capacity to process vast amounts of data. Attempts to mechanize were fairly slow because of the abundant supply of cheap labor, for the Post Office was notable in hiring women as clerks and telegraph operators. But from 1893 onward, efforts at mechanization did appear, with the introduction of automatic letter-folding machines, typewriters, calculating machines, and loose-leaf files.[38] Some parts of the organization were forced out of the city center by their size and the cost of land: the savings bank, for example, moved to Hammersmith, where the Office of Works put up an enormous building housing 3,200 officers and clerks in Blythe Road (1899–1903) just behind the new exhibition center at Olympia, thus helping to accelerate development in the suburbs.[39]

The senior engineers of the Post Office, in particular the chief engineer, were in positions to control innovation and propagate ideas about the need to adopt new technologies. Three chief engineers became presidents of the Institution of Electrical Engineers (IEE) between 1880 and 1905.[40] Particularly ubiquitous was William Preece (figure 3.1), who worked for the Post Office from 1870 to 1899 (serving as chief engineer from 1892 to 1899).

Besides serving on the Council of the Society of Arts and twice as president of the IEE, Preece was a Fellow of the Royal Society (1881), an advisor to the British Museum on lighting, and a prolific writer, lecturer, and publicist on telegraphy, electric light, and (after an initial dismissal of the technology) the telephone.[41] He took out eleven patents, gave evidence to Parliamentary enquiries and served on government committees, joined the Automobile Club (such membership served, in the club's early years, as a key indicator of technological modernity), wrote numerous reports on

FIGURE 3.1
London. Sir William Henry Preece, chief engineer of the General Post Office and promoter of electrical communication, by Beatrice Bright, 1899. Gazing thoughtfully at the viewer, Preece is depicted as successful and authoritative, the paper in his hand hinting at professional expertise. Nothing alludes to the dirty realities of electrical engineering. © Science & Society Picture Library.

his travels and attendance at congresses abroad, and ensured that the Post Office contributed to displays at many of the exhibitions in the metropolis. His scientific disputes with Oliver Heaviside, together with a tendency to be complacent about the superiority of Post Office methods, have perhaps obscured Preece's high-profile role as public servant and tireless publicist for the new technologies in metropolitan life.[42] For example, he contributed to the catalog of the 1892 Crystal Palace Electrical Exhibition, gave addresses at local municipal electrical exhibitions (in 1891 at St. Pancras Vestry Hall, and in 1894 to the Association of Municipal and Country Engineers on the Hampstead Central Electric Light Station), and even talked at the East End settlement Toynbee Hall in 1896 about wireless telegraphy, becoming the first to welcome Marconi's invention.[43]

Indeed, it was from the roof of the Post Office that Marconi made the first public wireless demonstration in Britain, and Preece's support marked a dramatic turning point in the young inventor's life.[44] Marconi arrived at Preece's office in July 1896 with an introduction from A. A. Campbell Swinton, another London-based consulting engineer and contractor. Marconi carried two large bags of apparatus, and demonstrated to Preece the ability to transmit a signal without wires. Excited by the invention, Preece put the expertise and facilities of the Post Office behind Marconi, allowing him to improve the equipment in the Post Office workshops, and organized a demonstration to Army and Navy authorities. Preece thus became Marconi's "first, and most potent, patron."[45] He spoke frequently about the exciting new developments, including to such elite audiences as the Royal Society and the Royal Institution. When Marconi formed a commercial company, however, he no longer found Preece's patronage necessary. Backed by family money, Marconi engaged an academic and consulting engineer, J. A. Fleming from University College, to help with experiments. Fleming was also well able to provide the sort of scientific support needed against theoretical rivals such as Oliver Lodge.[46] Nevertheless, the contribution of the Post Office workshops and engineers, together with Preece's public advocacy, were crucial in setting Marconi and his new technology on the path to success.[47]

By the turn of the century, the General Post Office was regarded as a notable feature of London, one which should be visited, and not simply for the purpose of posting a letter or picking up *poste-restante* items. It was featured in popular texts about electricity, such as Frith and Rawson's *Coil and Current,* which included "A Peep at the Post Office."[48] Much wonder was expressed at the immense size of the Telegraph Office, the marvelous accuracy of the instruments, the forest of batteries, and the astonishing arrival of foreign telegrams several hours before their apparent dispatch. Baedeker in 1908 recommended visiting the Telegraph Instrument Galleries ("admission by request from a banker or other well known citizen") containing five hundred instruments with their attendants and four steam engines that powered pneumatic tubes sending messages to other offices in the City or the Strand.[49] The equipment for operating the telegraphs and telephones was exhaustively described in the Institution of Electrical Engineers' *Electrical Handbook of London* (1906), which further noted that one wing had a "vacuum cleaning apparatus" for extracting dust, an emphatically modern innovation.[50] An institution which dealt with hundreds of thousands of transactions daily, the Post Office was thus the object of admiration, redolent of powerful technology but open to view by the interested public, massive in its symbolic and physical presence as the metropolitan center of global communications and imperial modernity. It is no wonder its senior scientific officers were highly respected and acted with confidence upon the metropolitan stage.

EMERGING METROPOLITAN ELITES: THE INSTITUTION OF ELECTRICAL ENGINEERS

A younger institution than the Post Office, the Institution of Electrical Engineers (IEE) was founded in 1871 as the Society of Telegraph Engineers. The IEE provides an opportunity to unpack the ethos and vision of a new technological profession and how it saw its place in the city. Its membership grew rapidly, with many joining from the Post Office telegraph services, the armed forces, and manufacturing; academic scientists such as William Ayrton, John Perry, and Sylvanus P. Thompson became members as well. The society maintained an office in Westminster, and for many years held its meetings in the Institution of Civil Engineers' capacious building.[51] In 1887 William Preece suggested a more "representative" name for what was now a national organization, and in 1889 the society became known by the more formal name Institution of Electrical Engineers. Relations were sometimes fraught between academic scientists and "practical men" engaged in the industry, as Preece's venomous debates with Oliver Heaviside showed.[52] Nor were the institution's forays into national politics always successful, particularly as those members working in the electrical supply industry were adamantly opposed to any proposals which would result in their enterprises being taken over by the LCC. This opposition helped to block the creation of a unified municipal electrical supply until the 1920s.[53] However, if we examine the IEE in the first decade of the twentieth century, we see an institution which acted as a common forum for academic engineers, industrialists, and civil scientists, was firmly embedded in the metropolis and keen to show its modernity, and whose leaders were ready to show leadership in international affairs.

The year 1906 was especially significant for the IEE as it marked the setting up of the International Electrotechnical Commission to establish standardized terms and ratings for electrical equipment, a move that had been discussed at the electrical congress which took place two years earlier at the 1904 St. Louis International Exhibition. Since London proved an eminently suitable location for the commission, its offices were established there, with the respected—if aged—Lord Kelvin elected as president, and Colonel Crompton as honorary secretary. Crompton, an electrical manufacturer and entrepreneur, would become president of the IEE for the second time in 1908, following Glazebrook (1906–1907) of the National Physical Laboratory, who was also very active in the IEE.[54] At the time that the commission was being established, an extended international gathering of electrical engineers took place, which provides a fine example of how London was perceived and projected by the metropolitan electrical engineering fraternity.

This view of London was conveyed in the *Electrical Handbook for London*, produced in 1906 by the IEE. At one level a paean to the glory of the Institution of Electrical Engineers and its work, the *Handbook* was primarily intended as a guide for visiting foreign electrical engineers during their stay in London. A map was included, showing steam and electric railways and tramways, open or in the process of construction, together with a number of sites of electrical interest marked in red (figure 3.2).[55]

A range of important and relevant buildings were identified—educational institutions, West End clubs, superior hotels, two museums and exhibitions, and the key government regulatory building (the Board of Trade), as well as electrical generating plants (privately and municipally owned), manufacturers of electrical equipment, and of course the General Post Office, the center of global communications. Here was modern

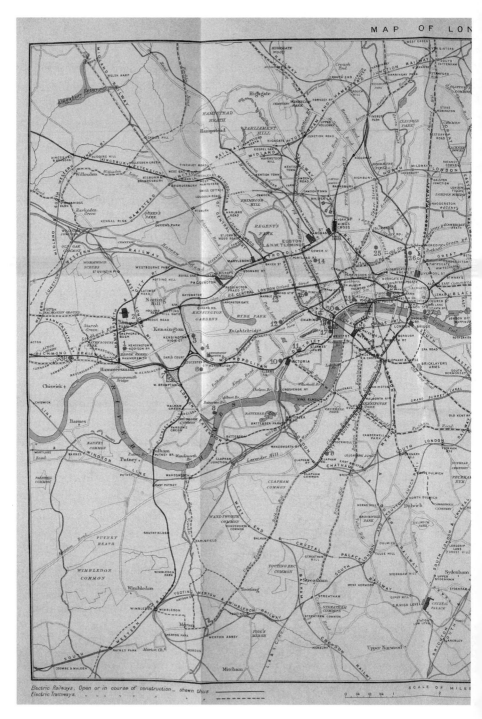

FIGURE 3.2

London. *The Electrical Handbook of London*, IEE, 1906, frontispiece. Modern London is revealed in this map of the principal electrical institutions and undertakings in London. Note that it included electrical underground lines in the process of

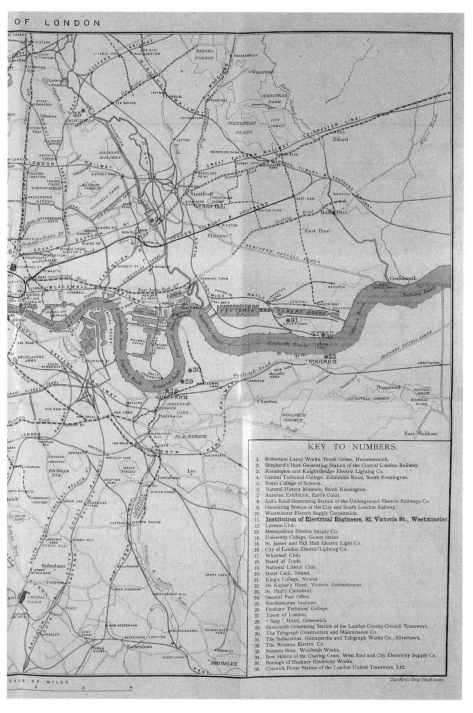

construction (heavy black lines) as well as those planned (heavy dashed black lines), emphasizing the dynamic and changing electrical character of the city. © The British Library. All rights reserved Ac4472.

London revealed. The text, lavishly illustrated with photographs, described in detail the electrical equipment to be found in each institution or plant and the layout of facilities, emphasizing the "exceptional interest" of particular systems or machinery.

Two further points illustrate how the IEE saw itself in the metropolis, and how modernity was demonstrated in the text. First, most of the institutions were described in terms of their precise urban context and their functional relation to that context. For example, the Northampton Institute (one of the new technical schools) was described as being "in the heart of the manufacturing district of the East End of London," and since this was the jewelry and clock-making district, every facility for instruction in these trades was provided, including a well-equipped electroplating laboratory and electrotyping shop, and the departments of technical optics and horology.[56] Secondly, most of the institutions were given historical context. The Royal Institution was described almost entirely through a quotation from Sylvanus P. Thompson's biography of Faraday. And it was noted that Johnsons, a manufacturer of cable-making and electrical generating machinery located on the outskirts of London in Kent, had suspended from the ceiling of their reception room a unique gasolier, made from the original grapnel used on the SS *Great Eastern* in recovering the transatlantic cable in 1865.[57] The past was thus recast in a way that emphasized links to the present, stressing continuity and progressive development, a pathway remembered and marked by the presence of great men and iconic objects.

The efficient management of the 1906 foreign visit both demonstrated the IEE's modernity and contributed to its effective establishment of international connections. Not surprisingly, a couple of years later, in 1908, the institution contemplated acquiring a building of its own, appropriate in size and stature to its status as an important national organization. After a search, a suitable site was located in Savoy Place on the Embankment. A pleasant, traditional, red-brick building, its interior was refitted in a bold modern style, complete with panels containing the great names of electrical science—Faraday, Volta, Ohm, Ampère, Kelvin, Hopkinson, Maxwell, and Henry.[58] Within this central location in the metropolis, there is a nice sense of the engineers putting themselves in the context of a long tradition of science while simultaneously showing an enormous enthusiasm for modernity.

The history of the IEE demonstrates how metropolitan elites established their legitimacy and authority to speak for a modern scientific industry.

Their acute awareness of global developments led them to see the need for the management and cooperation of an International Electrotechnical Commission. Such contacts helped both to cement brotherly feelings toward kindred engineers from overseas as well as to reinforce a sense of confidence in London as the center of empire.[59]

APPROPRIATE FORM FOR BUILDING THE MODERN CITY

The last decades of the nineteenth century saw extraordinarily rapid urban development. Many Londoners complained bitterly that familiar landmarks disappeared overnight and that the streets were always being dug up. The negative effects of this rapid development encouraged Sir John Lubbock to promote the first Ancient Monuments Act in 1882, a cause close to his heart.[60] However, the achievement of the Metropolitan Board of Works (the forerunner to the LCC) in developing a proper metropolitan drainage system, building new pumping stations, and creating the Thames Embankment should not be underestimated.[61] Some architectural developments were designed in the "grand manner" thought appropriate for a self-conscious imperial capital, particularly after two successive Jubilees and then the death of Queen Victoria in 1901.[62] As a movement, though, town planning mostly occurred outside London. Although the famous 1910 international Town Planning Conference and exhibition (attended by Chicago architect Daniel Burnham, among others) took place in the metropolis, it seems to have had little direct impact on the elites discussed here.

Besides the loss of municipal history, the public was also sometimes affected negatively by the expansion of scientific and technological infrastructure. In the 1890s, scientists W. E. Ayrton, Arthur Rucker, and Norman Lockyer helped fight off an extension of the underground railway that would have run underneath their laboratories in Exhibition Road, South Kensington, because they were afraid it would disturb their delicate electrical experiments—thus demonstrating that scientists were quite capable of selfishly defending their own interests to the detriment of the public good.[63] And London residents were forced to deal with a confusing hodgepodge of service providers. The national grid was not operational in Britain until the early 1930s, and in 1913 there were still twenty-eight different supply companies catering to different districts of London, some with more than one plant, some burning coal and some refuse, some private and some municipal undertakings, each with their own type of output (figure 3.3).[64]

FIGURE 3.3
London. *Wood Lane Power Station,* by Charles John Holmes, 1907. Holmes's painting of the power station, viewed over the rooftops from his house, vividly illustrates the intrusive reality of power generation in a rapidly developing residential neighborhood. The station supplied electricity to the Kensington area and was itself technologically advanced, the first station to provide high-voltage, three-phase generation and transmission. © Museum of London.

Power generation remained an intractable problem, generating numerous lawsuits on account of noise and nuisance, as well as making a significant visual impact. Architects designing such buildings could either conceal them or try to invoke a sense of wonder at the sheer power of the technology. Charles Stanley Peach (1858–1934), an architect who made power generation his specialty, provides examples of both approaches.[65] For the Duke Street substation located on the estate of the Duke of Westminster in a fashionable, residential district, Peach wrote that "every precaution should be taken to prevent transmission of mechanical vibration, or escape of sound beyond the building."[66] He therefore sank the main part of the works under a central "Italian garden" in the square, and placed baroque-style pavilions at either end, one containing the entrance to the station and access to the machinery, the other a garden shelter (figure 3.4).

The building and its purpose were thus completely concealed and disguised within a naturalized garden landscape in an architectural style appropriate for a duke. However, at this stage of the industry's development the

FIGURE 3.4
London. Electricity transformer station, Duke Street. Duke Street, part of the Duke of Westminster's estate, was in the smartest part of town. Hence, the architect felt that the appropriate form for this modern technology, in contrast to that in figure 3.3, was concealment within a baroque architectural shell. © Nicholas Breach/RIBA Library Photographs Collection.

buildings required were often so massive that elsewhere Peach tried to make a virtue of necessity, as in the case of the station at St. John's Wood.[67] His spectacular plan had six gigantic chimneys (only one was actually built) reaching into the sky above a "Wrenaissance" factory building.[68]

While much new construction in the capital during this period revealed a commitment to modernity and change, this was especially true of buildings for scientific and technical education. Chief among these was the new City and Guilds Central Institution in South Kensington, a building on the itineraries of numerous visiting educationists, which would become a key component of the newly founded Imperial College in 1907.[69] Aston Webb designed buildings in a suitably impressive style for the Royal School of Mines, though the most flamboyant building in South Kensington was the Imperial Institute (built 1887–1893), where chemist Frederick Abel became the first director. Numerous polytechnic institutions were established as a result of pressure from advocates of technical education (including the

Society of Arts), the efforts of the LCC Technical Education Committee under Sidney Webb (the Fabian social reformer and politician), support from the City Livery Companies through the City and Guilds Institute, and local support in many of the suburban boroughs. These institutions were housed in buildings that generally combined technical with social and sporting facilities, thus helping to create a focus for local identity in many suburbs, alongside new town halls for the reorganized metropolitan boroughs.[70] Such new buildings for the promotion of education, together with those that were explicitly technological in purpose and essential for modern city life, displayed in bricks and mortar the changing face of the metropolis.

PART III   MUSEUMS AND MODERNITY: FROM CULTURES OF THE PRESENT TO CULTURES OF THE PAST

I applied to a man who sells photographs of such edifices for pictures of the main buildings. He had none. "What, no photograph of the South Kensington Museum!" I exclaimed, with some impatience. "Why, sir," replied the man, mildly, "you see, the museum doesn't stand still long enough to be photographed."

—Moncure D. Conway, *Travels in South Kensington* (1882)

By 1900 the results of Victorian enthusiasm for museums could be seen all over the metropolis, intended above all to provide rational instruction and entertainment for the masses, as well as to create collections to advance knowledge and represent the nation. As integral parts of the cultural and social fabric of the city, museums were testament to the energy of the men who established them, and reflective of imperial reach in the richness and variety of collections they contained. The London County Council acquired its first museum as early as 1901, when the tea magnate Frederick Horniman handed over to it a newly erected building in Forest Hill, South London, to house his collections of natural history, anthropology, and musical instruments. However, despite the plethora of collections open to the public, there was no museum attached to the Royal Mint, as in Paris for example, or to the Post Office, where progressive technologies might be enshrined. The Post Office more than made up for this by its consistent contribution to all the major exhibitions of the period, as will be seen below. London also lacked a collection such as that at the Musée Social in

Paris, which brought together material and campaigns on issues of social economy. There was an attempt, notably in the Parkes Museum, to display scientific expertise that could influence and improve the lives of Londoners. But at South Kensington, by contrast, the technological collections attracted much criticism by the end of the century. Both institutions are now discussed.

THE PARKES MUSEUM OF HYGIENE: A GUIDE TO THE CONSTRUCTION OF HEALTHY LIVING

The Parkes Museum was established in 1876 to commemorate the work of Edmund Alexander Parkes (1819–1876), an early example of the civil scientist elite: a former professor of medicine at University College London, he was the director of the Army Medical School at Chatham, and a tireless advocate of sanitary improvement.[71] Prestigious support for the museum came from the royal family, some of the City Livery Companies, and distinguished representatives from society, politics, and science. Douglas Galton, newly retired from the Office of Works, threw himself into the task, corresponding with Edwin Chadwick and Florence Nightingale to obtain their support. The museum immediately attracted donations from institutions in London as well as from the U.S. Army's Hygienic Department in Washington, and opened to the public in 1879. It was provided with space within University College London, whose existing laboratory was considered sufficient for hygienic work, and was run by an executive committee of six professors. The museum's function was to display objects relating to the science and practical applications of hygiene, and to improve public awareness and knowledge through demonstrations and lectures, which were aimed initially at men in the building trades. Numerous exhibits explored the relevant sciences (bacteriology, chemistry, demography, and preventive medicine), but the bulk were concerned with construction, water supply and sewerage, heating, lighting, and ventilation. Another section was devoted to personal and domestic hygiene, and displayed items of clothing, food, beds and other furniture, gymnastic apparatus, and soaps and detergents.[72] All in all the museum provided a perfect demonstration of Victorian concerns with the sanitary state of the nation and the improving role of the latest scientific and technological knowledge.

The life of the Parkes Museum was not altogether smooth. Only three years later, in 1882, University College decided it needed the space and the museum had to move. Premises were found close by in Margaret Street, between Cavendish Square and the shopping area of Regent Street, which meant that the museum was effectively detached from the college, becoming a properly public museum rather than an adjunct to teaching with some public access. Sanitary reformers and their aristocratic patrons maintained their support. Galton remained a key figure, not surprisingly given his devotion to sanitary science.[73] In 1889 the Parkes Museum merged with the Sanitary Institute of Great Britain (set up in 1876), which had been founded to promote reforms and provide a focus for improving standards in practice. Professional practice and more general public educational concerns were thus brought together.

In its earlier years, the museum appears to have been reasonably well visited, for it extended its open hours from two hours in the afternoon to all day on weekdays. However, as a display of expertise applied to urban life, many exhibits had scant appeal beyond a purely professional or occupational interest.[74] As time went on, the museum experienced difficulties in adapting to rapid change, and its educational and cultural benefits did not match up to the reality of its aging exhibits. H. G. Wells wrote a devastatingly sarcastic review of the Parkes Museum in 1897, attacking its foodstuffs, "of a grey, scientific aspect, a hard, hoary antiquity," statistical diagrams, drainpipes, models of cells in Pentonville Prison, and cremation equipment, which was so attractive that it "must have such an added charm of neatness and brightness when alight, that one longs to lose a relative or so forthwith, for the mere pleasure of seeing [the equipment] in operation." As Wells summed up, the museum offered "valuable hints how to live, and suggests the best and tidiest way in which you can, when dead, dispose of your body."[75] Nevertheless, as the Sanitary Institute grew in national importance, the museum remained a useful adjunct, with student visits increasing to the extent that, in 1900, a building fund was set up to look for larger premises.[76] In 1909 the Sanitary Institute moved to a new home in Buckingham Palace Road, where the museum survived until the end of the 1950s. The Parkes Museum's long survival was proof of a modest success, and provides an example of mobilizing different types of elite support with aristocrats, academics, and professionals working together. But to flourish it needed to find a home within a relevant professional institution, and it was a professional rather than an academic elite that facilitated that transition.

THE SOUTH KENSINGTON MUSEUM, ITS CRITICS, AND THE CITY AS A MUSEUM OF MODERNITY

As is well known, the South Kensington Museum was established following the 1851 Great Exhibition as an industrial arts museum by a government keen to support science and technology for economic and political reasons.[77] The South Kensington site, over the course of the next fifty years, became an immensely important complex of museums, colleges, and training schools, which formed the apex of the British technical and scientific education system and was home to collections of national importance. The erection of a cathedral-like edifice for the Natural History Museum (1881), the completion of the building housing the Victoria and Albert Museum for the arts (1909), the building of the City and Guilds Central Institution (1880) for technical training, and finally the foundation of Imperial College of Science and Technology (1907), led to South Kensington becoming a magnet for elites, ambitious students, and the general public alike. However, due to the rapid growth in institutions, there were competing groups within those elites—the academics, their political masters, and the civil scientists appointed to museum positions. South Kensington might be well known and often admired abroad, but it was a highly contested site, particularly while the polemical T. H. Huxley, professor of biology in the government-built Science Schools, was active.

Huxley was, above all, concerned to make science, underpinned by agnostic ideas of Darwinian evolution, an integral part of education, as a powerful rival to classics at all levels. When his great rival, Richard Owen, finally retired from the Natural History Museum (just across the road from Huxley's College of Science), Huxley ensured that Owen's successor would be one of his disciples, W. H. Flower, thus eliminating one source of friction, and underlined his rejection of Owen's philosophy by arranging for a statue of Darwin to take pride of place in the museum's great hall.[78] However, Huxley was not interested in the scientific and technological collections at South Kensington beyond their function as examples of model educational apparatus or their possible use in advanced research.[79] As a result of this focus, which reflected the growing emphasis on laboratory teaching over collections-based knowledge, Huxley and his scientific colleagues, such as Edward Frankland (and later W. E. Ayrton and Henry Armstrong), were wholly occupied with establishing the colleges and laboratory facilities that would take their science forward.

The Department of Science and Art, the political paymaster of the South Kensington institutions and located within the same site, was also riven with internal rivalries and battles, particularly under Sir John Donnelly, head of the department from 1884.[80] Donnelly's main concern was to set up a workable system of technical education. He had long promoted the cause of technical education through the Society of Arts, where he was a close friend of Trueman Wood and served more or less continuously on the society's council from 1870 on. While Donnelly certainly spent time trying to improve the buildings on the South Kensington site, the role of collections was underdefined in comparison with that of teaching and education.

This lack of definition was due in large part to the way that the scientific collections had developed. At their foundation, the collections were intended to be of modern apparatus and machines, shown alongside the products of the most modern forms of manufacturing. In practice, the exhibits rapidly became out of date. The scientific exhibits, then, became simply another set of collections among a range that had accrued to the museum by gift, bequest, purchase, or negligence.[81] Nevertheless, an opportunity arose to clarify their purpose when the museum organized a large, high-profile exhibition, international in scope: the "Scientific Apparatus: Special Loan Collection 1876."

The exhibition was the result of one of the recommendations of the 1872 Devonshire Commission on scientific education, of which Norman Lockyer had been secretary.[82] It proved most successful, generating an enormous amount of involvement from everyone active in metropolitan science, and a capacious literature of catalogs, lectures, and reports in the press.[83] The lengthy catalog ran to two editions, and a *Handbook* was produced in the style of a South Kensington Museum science handbook, translated into French and German.[84] Reports of conferences and free evening lectures were delivered by notable figures such as the Earl of Rosse, John Tyndall, William Preece, and William Spottiswoode. These lectures were, arguably, what helped to define "scientific progress" as a series of iconic experiments conducted by great men, which could be understood over time and displayed by means of the scientific instrument.[85] A memorial urging the creation of a permanent physical science museum was addressed to the Lord President of the Council in July 1876, signed by J. D. Hooker, then president of the Royal Society, and by almost everybody who was anybody in British science.⁸⁶ It was not, however, successful in establishing a specialized museum.

This was for several reasons. The political momentum was not maintained, and Disraeli's government had many other causes for concern. The Department of Science and Art's budget was overspent, on the 1876 exhibition and other expenses, and a new institution would mean increased staff. Questions were raised in the press about who would constitute the audience for the proposed museum. And there remained some confusion about whether scientific apparatus would include new inventions, which added to the lack of clarity surrounding the status and future of the Patent Office Museum already housed in South Kensington, whose superintendent Bennet Woodcroft retired in March 1876.⁸⁷ Also, despite an impressive display of strength from the general scientific community, there was no single individual or group to drive forward the idea of a new and separate science museum.

With both relics of the past and the artifacts of modern science and technology continuing to find their way into the collections, the philosophy underlying the South Kensington Museum was not particularly clear. Certainly the museum was always acutely sensitive to examples elsewhere; there were constant references to the Conservatoire des Arts et Métiers in particular, from the 1850s through to the 1890s.⁸⁸

Meanwhile, increasing criticism of the South Kensington Museum began to come from different quarters.⁸⁹ Some critics felt that museums should concentrate wholly on the present and expunge as rapidly as possible all trace of the past. Some felt that the collections on the site were simply too disparate, huge, and unmanageable. The economist W. S. Jevons ridiculed the idea of civilizing the masses by showing them glass cases filled with beautiful objects illuminated by electric light. In his essay "The Use and Abuse of Museums," he homed in on South Kensington with its interminable galleries and confusing layout.⁹⁰ He asserted that a proper technical museum would have to be enormous to cover everything, its contents would become obsolete within ten years, and it would be more like a shop trying to stock the latest models. Instruments underwent such frequent modification that in any case the best place to see them *was* in a shop.⁹¹ Writers such as H. G. Wells—who called it "a fungoid assemblage of buildings"—were only too happy to lampoon South Kensington.⁹²

In 1897–1898, in a highly critical Committee of Enquiry, officials (Donnelly in particular) received a mauling; as a result, the Department of Science and Art was amalgamated with the Board of Education, and the division of the science and art collections was initiated. It would seem that South Kensington had been rationalized and the problems of the museum so obvious in the 1890s had been resolved.[93] But while the new building for the arts went ahead and was finally opened in 1909 as the Victoria and Albert Museum, the proposed Science Museum failed to get through the political and financial logjam, and building plans remained stalled for two more decades.[94] The only area of agreement was that the former museums had lacked space and thus been overcrowded, and the science collections had been difficult to look after properly.

Nonetheless, there was something of a mismatch between the ferocity of public criticism in the 1890s and the actual situation on the ground. This has helped to shape a story of the "struggle" to establish the Science Museum which underlies many histories.[95] Members of the scientific elite did rally to the cause of a properly organized science collection, but they were in general not those working in South Kensington. Supporters included Lyon Playfair, the aged and ennobled chemist and MP, and Alexander Geikie, Scottish geologist and president of the Royal Society (1908–1913).[96] In the first decades of the twentieth century, Glazebrook took an active role in promoting the science collections. As a governor of the new Imperial College and an 1851 Commissioner (the Commission controlled the real estate), he became a key figure in South Kensington as well as in larger metropolitan affairs. The first block of the new Science Museum was eventually officially opened in 1928, though the main center block was not completed until 1962.

By the turn of the century, London itself was seen as the museum of modernity, and was often recommended as the best place to view the latest developments, as Miriam Levin has noted of Paris.[97] Journals such as the *Illustrated London News* regularly illustrated new developments—underground railways, electric tramcars, Jablochkoff lamps on the Embankment, and electric light at the British Museum.[98] In any case, the primacy of private enterprise meant that the City "operated as something of a laboratory" for innovations such as electric lighting.[99] Jevons wrote that a busy factory was one of the very best kinds of educational museum, and recommended that more factory proprietors should open their doors to visitors.

While manufacturers might demur on account of the dangers of revealing trade secrets, there were no such obstacles to viewing government works of all kinds. Entrance was easily gained to facilities such as the Royal Mint, the Dockyards, and Woolwich Arsenal, which suggests a new relationship of trust between government institutions and the public. Turn-of-the-century guidebooks recommended institutions and works to visit, and described how to get tickets or introductions, as noted above in the case of the Post Office. Baedeker, for example, recommended visiting Siemens Brothers in Woolwich, where submarine cables were made.[100]

If the city itself was a showcase of modern technology, then museums stood at the intersection of present and past. Even Wells could not resist using South Kensington in *The Time Machine*, as the Palace of Green Porcelain—the museum as time machine, preserving amid universal decay the machines and objects of the once great civilization of man.[101] As collections inexorably increased, it became natural to frame the past as the history of progress, often personified in terms of discoverers and discoveries. This view of the past was an attractive counterpoint to modernity: old, but relating to the present; both time-bound and timeless, but increasingly a contrast to the exciting world of exhibitions, where the public might expect to find displays that emphatically related to modernity and the future.

PART IV   EXHIBITING PAST, PRESENT, AND FUTURE

[There are] two types of mind. ... The former type of mind, when one gets it in its purity, is retrospective in habit, and it interprets the things of the present, and gives value to this and denies it to that, entirely with relation to the past. The latter type of mind is constructive in habit, it interprets the things of the present and gives value to this or that, entirely in relation to things designed or foreseen.

—H. G. Wells, "The Discovery of the Future" (1902)[102]

Exhibitions provided a stage which brought the various elites of the city together: Trueman Wood and Lockyer were prominent, for example, and William Preece boasted in 1907 that he had been involved as exhibitor, juryman, committee member, or commissioner with fifteen universal expositions overseas and a similar number of major ones in Britain, together with many smaller ones.[103] Expositions were sites where, as Bernhard Rieger says, "contemporaries had to find their bearings in a maze of modern time."[104] During the period leading to World War I, there were no universal expositions in London comparable with those held in Paris in 1889 and 1900, or in Chicago in 1893. There were, however, numerous exhibitions which were "international" in character, in that exhibits from other countries were shown and foreign visitors encouraged to visit. London's spatial organization and transport infrastructure had a critical impact on defining the location and style of exhibition sites. Spectacles of science and technologies that were provided for the edification of fairgoing visitors certainly depicted stories of progress, but fewer than might be expected were applicable to London's contemporary problems. Nevertheless, the old and new were juxtaposed in fresh ways to provide continuities between past and present. And ideas about the future were further promoted through entertainments and displays of "technological fun."

SPATIAL ORGANIZATION AND THE SHIFT TO THE SUBURBS

Shortly after the 1851 Exhibition closed, discussions began about what to do with the Crystal Palace. The southern suburbs were alluring, with room for development, space for a park and appropriate kinds of rational entertainment, and possibilities for excellent transport links. The Palace was thus moved to Sydenham, reopening in 1854.[105] This marked the first stage in shifting exhibition sites out of central London. Two further sites were constructed in Islington and Muswell Hill to the north of the city.[106] All three sites were run by private companies, and took advantage of—or built—railway links.

South Kensington had been a suburb at the time of the Great Exhibition in 1851. Although schools and museums rapidly filled up the site, exhibitions continued to be staged in the arcades and gardens to the rear, with buildings such as the Albert Hall pressed into service if needed. Four "international exhibitions" were held there from 1871 through 1874, each one focused on a restricted group of manufactures rather than being truly universal. Attendance declined, and they ceased to be held until the 1880s, when four more international exhibitions, each with a thematic focus, were organized: Fisheries (1883), Health (1884), Inventions (1885), and Colonial and Indian (1886).[107] These were successful in attracting visitors, but expensive to mount. After the last of these, such exhibitions on the South Kensington site ceased, as the museums and schools won the competition for the space.

International exhibitions spurred the second stage of suburban development, from the 1880s on. Two new venues were developed as commercial ventures to compete with Islington and Sydenham. The first was Earls Court, to the west of South Kensington, an area rapidly being built up. It was an awkwardly shaped site, which had arisen out of the chaotic development of London's rail network.[108] The promoter, John Whitley, saw the possibilities of creating a different format, with an open-air arena on one part of the site, a standard exhibition building in the center, and gardens and amusement grounds on the third portion. He mounted a series of exhibitions that were part trade show and part exotic entertainment, starting with the fabulously successful American Exhibition (1887).[109] This started a pattern of exhibitions, focused on one country or theme, which would continue until World War I. Just to the north of Earls Court was its rival site, Olympia, which opened in 1886 with a series of circuses and spectacles of the most extravagant kind.[110] But it had an uncertain career, going through a series of owners until around 1905, when it settled upon a program of regular trade shows (including the annual Motor Show) and exhibitions.

Olympia brought the Hungarian-born impresario Imre Kiralfy (1845–1919) onto the London stage. Kiralfy had pursued his career first in Paris and then in the United States, where he had created the spectacle "America" for the Chicago Auditorium Theater, coinciding with the 1893 Chicago exposition. He was so impressed with this exposition that he first built the Earls Court exhibition grounds as a small-scale version of those at the Chicago exposition. When he had outgrown that venue, Kiralfy took his admiration a stage further in 1908 with the development, in Shepherd's Bush, of White City, named in homage to the Chicago site.[111] He cultivated political contacts, particularly through becoming a member of the British Empire League, which gave him an entrée to high-level politicians and statesmen such as the eighth Duke of Devonshire.[112] These connections enabled him to undertake the organization of the officially sponsored 1908 Franco-British Exhibition, designed to seal the 1904 *entente cordiale*, and then the 1910 Japan-British exhibition celebrating the Anglo-Japanese alliance (figure 3.5).

The 1908 Franco-British Exhibition was the first time a commercial site was used for an official exhibition, and underlines the impact of international exhibitions such as the one in Chicago in 1893. The last suburban exhibition site to be built was Wembley Stadium in northwest London in

FIGURE 3.5
London. Poster for the Japan-British Exhibition, 1910. The poster uses up-to-date graphic style and typography while emphasizing the traditional roots of British and Japanese culture in "Britannia" and the kimono-clad woman. The White City exhibition site is depicted as an exotic wonderland of imaginative architecture and a nighttime fairyland of electric lighting. © Museum of London.

1922, a combined athletics and entertainment center that provided an up-to-date venue for the massive British Empire Exhibition staged in 1924–1925.[113] Thus London became ringed by suburban exhibition sites that provided space for leisure, entertainment, and ever-grander spectacle. All were dependent upon or owed their existences to the railway, and all were run by private companies or entrepreneurs.

SPECTACLES OF URBAN PROGRESS AND OLD LONDON
London exhibitions may not have been truly *universelle*, but they were deeply influenced by the genre.[114] Exhibition organizers, especially Kiralfy, complained that the London County Council (LCC) gave no financial support, unlike corresponding agencies in Paris.[115] Such activities were more

likely to be carried out by the London Chamber of Commerce, itself only founded in 1881.[116] The LCC's only role in expositions had been to express concern about municipal regulations for traffic, health, and safety. This changed, however, in 1910, on the occasion of the Japan-British Exhibition at White City. Invited to provide an exhibit, the LCC decided that its exhibit should be "worthy of the Council's position as the greatest municipal authority in the world," and voted for the relatively generous sum of £1,000 toward the cost of preparation.[117] This was belated recognition that exhibitions could serve as a means of promoting a vision of the city.

Nevertheless, exhibitions were prime sites for creating displays and spectacles that related directly to London and its environment. Their effects were not necessarily visible in new plans or changed projects for the city (as was the case in Chicago) but influenced public attitudes toward the city, as well as allowing campaigners platforms to promote particular ideologies and projects. Exhibits provided examples of the latest developments in lighting and motive power, sanitation, and street cleaning. Such exhibits were most notably displayed at the 1884 International Health Exhibition at South Kensington.

This exhibition brought together many from elite networks. Sir John Lubbock, Douglas Galton, and Frederick Abel served on the executive committee, along with the Lord Mayor of London, various aristocrats, MPs, and medical men.[118] The Health Exhibition was a testament to concerns about public health and interest in the field of sanitary science. Its international component was rather limited—including some publications from Japan, various exhibits from Belgium, and an Indian section—but the displays of heating, cooking, food preservation, rational dress, water, and lighting served as a demonstration of progress in sanitary science and a reminder of the recent restructuring of London's sewer system. The eight metropolitan water companies seized the opportunity to promote their wares by joining together in a "handsome pavilion" with a large central fountain surrounded by eight drinking fountains, sections of filtering beds, and models of the companies' works.[119] The message was clear—London had clean and pure water, though it was provided by eight private companies. The publications issued by the exhibition organizers (in particular, Trueman Wood at the Society of Arts helped to ensure that so much information was published) reiterated the message about improving urban health. One, titled *Health in Relation to Civic Life*, contained articles on fires and fire brigades (with much emphasis on new machines and up-to-date firefighting techniques), ambulances, street cleaning, and legal obligations with respect to dwellings of the poor.[120] Trueman Wood and Abel contributed a long piece on water supply and distribution to another volume.[121]

In general, later exhibitions were upbeat about measures for tackling London's problems and those of other big cities. The Japan-British Exhibition of 1910 featured models of Tokyo and Osaka, provided by the municipal government of those cities, whose waterworks were apparently much admired.[122] The officially sponsored LCC exhibit covered public health, education, fire protection, traffic, and weights and measures. The traffic section concentrated on tramways and street improvements, displaying the most recently designed electric tramcar. It also included a large model of the most recently constructed thoroughfare, the Kingsway and Aldwych scheme, which remains the sole Haussmann-inspired piece of London development.[123] With its subway for electric trams and an underground railway, the LCC was justly proud of this example of advanced urban development, and juxtaposed the model with a cartoon of the "network of mean streets" swept away by the new development.[124]

While the future as such was not displayed, it was implied by the way that the present was articulated and put on show.[125] Models of towns and of particular schemes were regarded as prize draws. For the 1887 American Exhibition in London, George Pullman had offered to send a model of his planned city of Pullman (a company town development); instead, an enormous model of the new City Hall in Philadelphia was shown, though it was somewhat damaged in transit.[126] At the Franco-British Exhibition (1908), a Pavilion for the City of Paris was erected. As the enthusiastic writer of the *Illustrated Review* of the exhibition wrote, the pavilion's display demonstrated the care the Parisian authorities took "to improve the welfare of each of her citizens by the diffusion in every home of light, of water, of electric power, animating their machinery and carrying the burden of human thought" (by which the writer probably meant the telephone).[127] Public officials, the writer went on, should not be made fun of or satirized, given their labors in a host of spheres, from education and hygiene to laboratories for materials testing and the analysis of foods, to the suppression of murder, theft, arson, and other crimes. This list nicely encapsulates the way that such displays worked. The horrors of the past were invoked by the specter of murder and arson, while the peaceable present of properly protected citizens was assured by the ceaseless efforts of the toxicology laboratory, the anthropometric service, the police forces, guardians of the peace, commissioners of morals, and so on.[128] Science in Paris was explicitly personified in the expert public servant, and order in society could thus be assured.

London was much slower to claim the same authority for its public services or officials and to take advantage of such opportunities to publicize their work.[129] In London exhibitions generally, there was a comparative neglect of social questions, though such topics were covered in British pavilions in overseas expositions. Patrick Geddes (1854–1932), the Scottish thinker on social evolution and city planning, was absent from the major London exhibitions discussed above, and held his annual summer meetings on the study of evolution in society in Edinburgh, except for 1900, when he set up shop at the Paris exposition. This brought him an international reputation. Indeed, when Geddes came to plan his "Cities and Town Planning Exhibition" for the first international Town Planning Conference, held in London in 1910, and while working on his book *Cities in Evolution*, it was to Paris, Vienna, Washington, and the Burnham Plan for Chicago that he looked, rather than to London.[130] Nevertheless, a sense of "order" was arguably a general feature of scientific displays, as embodied, for example, in the regular exhibits by the National Physical Laboratory (NPL). These emphasized the laboratory's role in precise measurement and setting standards, and therefore in helping to sort out the confusions or irregularities of a complex industrial world, both now and in the future.[131] Glazebrook ensured that the NPL was always represented at exhibitions, thus bringing it to the attention of a much wider public and enhancing its authority.

The sense of progress from the past to the present, and into a bright future, was further heightened by the deliberate use of historical contrast. This motif was used in individual exhibits, such as the Post Office's entry, which under Preece and his successors had a regular display of historic and modern equipment showing the progress of telegraphy, and was sent to exhibitions at home and abroad.[132] But in the story of progress, London came into its own with the creation of "Old London." An "Old London Street" was first created for the 1884 Health Exhibition. This was an imaginative grouping of full-scale replicas of old houses, almost all from the City of London, grouped to provide an impression of the city as it had existed before its destruction in the Great Fire of 1666. It had an overall articulating function of contrasting the mistakes of the past (dirty, overcrowded, ill-lit houses) with the progress and innovation of the present (new sewer systems, slum clearance, and clean water).[133] However, while some visitors might have recoiled from the portrayal of the negative aspects of Old London living, the display rapidly became an enormous success as a quaint and picturesque attraction.

As a result, the "Old London Street" was retained the following year for the International Inventions Exhibition, and was even improved by the replacement of its boarded flooring with a more "authentic" pavement and gutter.[134] Its ostensible purpose shifted from showing the unhealthy city past to providing a place where "useful comparisons between ancient and modern handicrafts" could be drawn, though the organizers as good as admitted that the display was retained primarily as an attraction, since its connection to the theme of inventions was tenuous at best (and the irony of its inclusion was lost on contemporary commentators).[135] Yet the Street did articulate the overall plan of providing a contrast between the old and the new, with the latter being embodied by the power-generating room. As the *Illustrated London News* put it, the visitor stepped "from the old world into the new—from the ancient London street into the engine-room where the electric-lighting dynamos are busily working, generating electricity for the incomparably magnificent and extensive system of lighting which charms the eyes of all beholders at night."[136]

At the same time, the exhibit tapped into concerns about the disappearance of much of old London, a concern that had led to the foundation of the Society for the Protection of Ancient Buildings (1877) and the Society for Photographing Relics of Old London (1879) and that underlay Lubbock's interest in historic preservation.[137] Numerous commentators lamented the disappearance of old landmarks, though the urge to preserve did not necessarily imply only a nostalgic longing for the past.[138] In exhibitions, Old London (and its imitators) continued to function as a positive contrast to the present that could be enjoyed simply for its picturesque qualities, while providing a reassuring sense of continuity between past and present. It supported a story of progress, full of interesting incident, but in the end confirming the essential unity of past, present, and future. Old London could be found again (in slightly different form) in the Franco-British and Japan-British Exhibitions; and fairs in other towns (Edinburgh in 1886, Manchester in 1887, Berlin in 1896, and Paris in 1900) featured similar exhibits.[139]

FUN AND THE FUTURE AT THE FAIR

There was another way that the future was characterized in these exhibitions—it would be a future of fun, enjoyment, and broad participation. Amusements and diversions were of course essential for attracting as large an audience as the organizers could persuade to pay the entrance charge.[140] As one British commentator wrote, because Paris was the "pleasure capital" as well as the business showroom of the world, it was the only city where the success of a world's fair could be confidently assured.[141] Increasingly, to compete with Parisian achievements, London exhibitions employed modern technologies to create both the aesthetics and physical experience of new and exciting pleasures.

Not surprisingly, electricity was a key means by which the pleasures of the present and promise of the future were demonstrated, since it was thoroughly modern. It was adopted first in the South Kensington Fisheries Exhibition of 1883.[142] Vast sums were expended on electric lighting, especially in creating nighttime spectacles of illuminated wonders. The 1883 Fisheries Exhibition spent £10,397; a year later, the Health Exhibition spent £30,751, and for the Inventions Exhibition the figure rose to £53,515.[143] Magazines such as the *Illustrated London News* carried dramatic double-page spreads of the illuminated fairylands created.[144] Similar spectacles—"A Dream City: The Court of Honour illuminated with myriads of electric lights," for example—were still highly regarded more than twenty years later at the Franco-British and Japan-British Exhibitions.[145]

The attraction of such electric spectacles was, at one level, their novelty and the way that they turned a part of the city into an "enchanted ground" for a brief period.[146] Electric light allowed visiting hours to be lengthened and exposed visitors to spectacular landscapes of modernity, as the ordinarily sedate *Nature* put it in 1885:

Where before all was darkness, there is a scene of bewildering enchantment: fountains play and throw up into the air, now high, now low, solid sheets of illuminated water. ... When the silver light of the electric arc alone illuminates the fountains, broken by some magic power below into waterdrops, all the prismatic colours of the rainbow are observable, and, revelling in the beauty, one wonders how it is all brought about.[147]

This enchantment had an additional function for the exhibitors: it glossed over differences in electrical supply, its periodic unreliability, and well-reported accidents. The public needed to be persuaded to use electricity, first for light, and—by the early decades of the twentieth century—for cooking and heating. Exhibitions provided dramatic, but safe, demonstrations that the future was electric.[148]

By the time of the 1908 and 1910 exhibitions, a wider variety of sensations were offered to visitors. At White City there were journeys in "chariots and cars ... by earth and air and water," the giant "Flip-Flap," a scenic railway which provided a thrilling switchback journey, a "Canadian" toboggan run, and a swerving and jolting Irish jaunting-car (the "wiggle-woggle").[149] Cinemas were just beginning to appear in London, including the aptly named Electric Palace in North Kensington, built in 1909.[150] The new film technology also began to appear in support of particular exhibits—railway companies were notable for using film to show the delights of travel.[151] Visitors to fairs were given a wide choice of exciting sensations, quite different from those formerly provided in science and machinery demonstrations. Moreover, these experiences occurred in sites where the aesthetics of the buildings and the environment were as much the subject of comment as the exhibits themselves, and the organizers took the greatest care to exclude any reminders of a less attractive environment outside.[152]

Finally, we should remember that in Britain technological modernity was not at odds with the celebration of imperialism. London was unquestionably a world city, and exhibitions helped to reinforce the view of it as imperial capital.[153] Amid the Oriental-style architecture and electric entertainments of a site such as White City, technology, modernity, and imperialism found a happy meeting place.

PART V CONCLUSION

Ideas of change in London naturally varied enormously during the late Victorian period. The city was never wholly a Babylon, nor had it become, by 1914, a "White City." Pessimistic critics focused on the physical destruction of the old city caused by projects for metropolitan improvement. For them, change and progress meant the collapse and loss of much of the familiar landscape.[154] But, for every critic of the great city, there were others who found in London a modern, liberating, and stimulating environment.

This chapter has suggested that it was different groups and types of scientists and technologists who influenced the shape of development in London. We have also argued that corporate structures and institutions were as important in this period in defining the status and role of elites as were the individual educational or professional backgrounds of these men. Independent men of science such as Sir John Lubbock—the epitome of a traditional "social leader" directly involved with both urban governance and a multitude of metropolitan organizations—were increasingly the exception. Among civil scientists based in corporate organizations, there were a striking number of men with engineering backgrounds. William Preece was a professional specialist who vigorously promoted the vision of technological modernity epitomized by the Post Office with its imperial communications network. Richard Glazebrook followed a recognizably modern career path, pursuing academic work at Cambridge and other universities before moving into a public scientific organization. And men such as Trueman Wood turned organizations such as the Society of Arts to influential use. The diversity of urban scientific elites in this period should be emphasized, with no sharp division made between academic elites, civil scientists, and men such as Trueman Wood, Frederick Bramwell, and Norman Lockyer, who bridged different organizations and networks. Such men mixed freely across various organizations, moving between public and private enterprise and working alongside professional engineers, industrialists, academic research scientists, liverymen from the ancient City guilds, and consultants to the LCC; they also served on the governing bodies of Imperial College and London University, and organized learned societies and London clubs. The associational culture of an earlier period still provided an effective social glue and a communications network that enabled elites to function in the vast metropolis.

These elites did not necessarily have a consistent approach to solving the problems of the city in the new modern age. Lubbock, given his opposition to the creation of unified municipal services, certainly cannot be considered a radical; he also opposed the reform of London University (which Richard

Burdon Haldane eventually carried through). His presence on an organizing committee lent gravitas and became in time more ornamental than executive, though he never missed the opportunity for a good speech and used the openings of the Franco-British and Japan-British exhibitions to argue for international cooperation based on free trade.[155] Preece was on the council of King's College London, but resigned in 1894 in opposition to the dropping of religious tests for tutors and prospective candidates, which also put him at odds with moves to reform London University.[156]

At the start of this chapter, certain themes were emphasized as characteristic of London's culture of change: enthusiasm for technological modernity achieved in a progressive evolutionary manner, the key role of private enterprise and the importance of local autonomy, and a devotion to historic continuity. In this transitional period, the authority of elites was certainly fostered by the fact that London was a hugely important center for international communications. The city's organizations could also demonstrate their technological modernity while simultaneously positioning themselves within an age-old tradition, thus showing a natural progressive development rather than any revolutionary rupture. The Post Office managed this well, as did the Institution of Electrical Engineers. The major role taken by private enterprise, however, also indicated an acceptance of vested interests, local control, and a suspicion of many areas of centralized systems. Thus, London might appear backward and fragmented in its provision of services. However, as Chris Otter argues, this tendency was underpinned by a liberal political outlook and a "sensitivity to the local, and … resistance to system" extended to the world of electrical engineering. We should interpret this, Otter says, as the positive shaping of a distinctively British path, "characterized by cautiousness, eclecticism, and localism."[157] For example, the Post Office might have been monolithic and centralized, but electric lighting and urban transport remained diverse and locally provided until after World War I. A lack of standardization put locality first and the whole city second, and implicitly placed less emphasis on modernity.[158] The size of the city and its long history of localism and fractured governance encouraged such attitudes. But at the same time, the extension of the city into the suburbs—which the railways and new exhibition sites encouraged—allowed organizations to develop in a highly competitive and lively manner.

As mentioned earlier, there was an acute awareness among elites of developments taking place in other cities, whether in the extension of city services, new power station design, the establishment of museums and exhibitions, or the need to establish international standards relating to science and technology. Sometimes such comparisons were ignored (as in town planning, in many respects); other times, they were imitated (the construction of White City in a London suburb) or regarded with hostility (by utilities keen to maintain their local franchises and opposed to the modernity of integrated systems). Nonetheless, whatever the pressures of emulation or competition, at the same time the majority of Londoners felt that their city remained supreme. Given the continuing trust in free trade, the visible evidence of London's centrality as a global port and market, and the confidence of professional elites, comparisons with foreign places and institutions were usually exploited to exert pressure in local causes, rather than arising from any inner anxiety that London was falling behind. Comparisons with Charlottenburg were useful in pressing for support for technical education, but there was no appetite to stage a full-scale *exposition universelle* to rival the ones in Paris. London was a magnificent site for the display of scientific and technological modernity, and as such it was well used by metropolitan elites; however, it was not often used to show the application of science to the urban environment, the Parkes Museum being a notable exception.

Finally, an attachment to historic continuity was characteristic of London. A past was constructed by forward-looking institutions in order to demonstrate that modernity had been created through gradual change rather than revolutionary rupture. It became customary for these institutions to look backward and place themselves within a progressive historical perspective, as did the IEE. In museums, technological collections naturally lent themselves to being placed within a story of progressive development. A similar approach may be seen in exhibitions, where the idea of London functioned at different levels. On the one hand, it was a subject of display where the latest developments in urban infrastructure and management were exhibited. These displays demonstrated London's role as both the biggest municipal authority in the world and as an imperial capital. Modernity and imperialism could be happily combined in schemes for under-river tunnels or new developments such as the Kingsway-Aldwych scheme. On the other hand, the "Old London" exhibit notably functioned as a reassuring reminder that the familiar landmarks of the city could still be cherished in the brave new world of electric light, sanitary housing, tramcars, and railways. These implied continuities meant that the London on display was Janus-faced. The beauty of a picturesque past, the strength of the imperial present, and the modernity of a prosperous future could all be embraced at the same time.

ACKNOWLEDGMENTS

Particular thanks are due to my colleagues in this project, and to Ruth Barton, Robert Bud, Barry Doyle, Graeme Gooday, and Mark Patton, as well as those named in the various endnotes.

NOTES

1. Ably explored in Lynda Nead, *Victorian Babylon* (New Haven: Yale University Press, 2000). See also Richard Dennis, "'Babylonian Flats' in Victorian and Edwardian London," *London Journal* 33, no. 3 (November 2008): 233–247; and Andrew Lees, *Cities Perceived: Urban Society in European and American Thought, 1820–1940* (Manchester: Manchester University Press, 1985), 6.

2. It depends on what area is counted as London, but in 1871 in the area of London (as defined by the 1888 London Government Act) the population was 3.26 million; in 1901, 4.536 million; and in 1931, 4.397 million. During the same period, the population of Greater London more than doubled to 8.110 million.

3. Iwan Morus, Simon Schaffer, and Jim Secord, "Scientific London," in Celina Fox, ed., *London: World City* (London: Museum of London, 1992), 111–142.

4. Richard Dennis, "Modern London," in Martin Daunton, ed., *The Cambridge Urban History of Britain*, vol. 3, *1840–1950* (Cambridge: Cambridge University Press, 2000), 98, 121–125.

5. See Gareth Stedman Jones, *Outcast London: A Study in the Relationship between Classes in Victorian Society* (Oxford: Clarendon Press, 1971). Derek Fraser, *Evolution of the British Welfare State*, 3rd ed. (Basingstoke: Palgrave Macmillan, 2003), remains the standard work on the development of social welfare.

6. There was certainly anxiety about revolutionary strife, particularly following the example of the Paris Commune in 1871, but it never became dominant. The *Times* ridiculed a demonstration in London in support of the Paris Commune and reckoned that in London "the only fault of our metropolitan self-government is that there is rather too much of it." *Times* (London), 18 April 1871, 9, issue 27040, col. E.

7. Albert Shaw, *Municipal Government in Great Britain* (London: T. Fisher Unwin, 1895), 263. By the end of the century, developments in other British cities were increasingly held up as models, for example, Glasgow's trams and Birmingham's municipal socialism as developed under Joseph Chamberlain.

8. There were equally many reactionary voices such as those of John Ruskin, who was hostile to industrialization, or William Morris, who had a romantic vision of a craftsmen's utopia. While the writings of such luminaries were widely read and admired, including among elites, their impact in the practical world should not be overestimated.

9. For a discussion of whether people viewed this era as a period of steady progress or as a complete rupture from the past, see Martin Daunton and Bernhard Rieger, eds., *Meanings of Modernity: Britain from the Late-Victorian Era to World War II* (Oxford: Berg, 2001), 1–21.

10. Daniel T. Rodgers, *Atlantic Crossings: Social Politics in a Progressive Age* (Cambridge, MA: Belknap Press, 1998); Pierre-Yves Saunier, "Connections: A Municipal Contribution," *Contemporary European History* 11, no. 4 (2002): 507–527.

11. Quoted in James Boswell, *The Life of Samuel Johnson*, vol. 1, ed. with notes by Roger Ingpen (London: Pitman & Sons, 1907), 258.

12. W. D. Rubinstein, *Elites and the Wealthy in Modern Britain* (Brighton: Harvester Press, 1987). See also his "Britain's Elites in the Interwar Period, 1918–39," in Alan Kidd and David Nicholls, eds., *The Making of the British Middle Class?* (Stroud: Sutton Publishing, 1998), 186–202. Rubinstein argues that the landed and financial elites merged into a national elite, largely benefiting from interwar economic trends, and that there was considerable continuity in symbols and institutions.

13. Earlier, there seems to have been considerable integration in London between commercial elites and men of science, to take J. P. Gassiot, William Spottiswoode, William de la Rue, and Sir John Evans as examples. The commercial and scientific elites expanded and diversified with the emergence of new technological industries, but the close connections were not in themselves a wholly new phenomenon.

14. Robert J. Morris, "Governance: Two Centuries of Urban Growth," in Robert J. Morris and Richard H. Trainor, eds., *Urban Governance: Britain and Beyond since 1750* (Aldershot: Ashgate, 2000), 3.

15. The function of elites here is distinguished from the general activities of elite scientific figures in the public arena, as analyzed by Frank M. Turner, "Public Science in Britain, 1880–1919," *Isis* 71 (1980): 589–608. While it has clear resonances with Turner's analysis, the present paper identifies elites and their activities in relation to the city, not in terms of "public science," which Turner sees as a body of rhetoric, argument, and polemic used to justify their activities and to improve their position in the public arena.

16. Andrew Saint, ed., *Politics and the People of London: The London County Council 1880–1965* (London: Hambledon Press, 1989), provides a clear guide to the development of London's municipal structures.

17. The role of the expert in governance was to be trusted by the public to make the right decisions; see Christopher Hamlin, *A Science of Impurity: Water Analysis in Nineteenth Century Britain* (Bristol: Adam Hilger, 1990), especially chap. 6, 152–177.

18. Ruth Barton, "'An Influential Set of Chaps': The X-Club and Royal Society Politics, 1864–85," *British Journal for the History of Science* 23 (1990): 53–81; J. Vernon Jensen, "The X Club: Fraternity of Victorian Scientists," *British Journal for the History of Science* 5 (1970): 63–72; Roy MacLeod, "The X-Club: A Social Network of Science in Late-Victorian England," *Notes and Records of the Royal Society of London* 24 (1970): 305–322.

19. Mark Patton, *Science, Politics and Business in the Work of Sir John Lubbock: A Man of Universal Mind* (Aldershot: Ashgate, 2007). Lubbock's role in the university is exhaustively discussed in F. M. G. Wilson, *The University of London, 1858–1900: The Politics of Senate and Convocation* (Woodbridge: Boydell Press, 2004).

20. Obituary notice, *Proceedings of the Royal Society of London*, Series B 87 (1913–1914): ii. See also *Oxford Dictionary of National Biography* (Oxford: Oxford University Press, 2004); hereafter cited as *Oxford DNB*.

21. The term "liberal" still carried a sense of progress and outward-looking development, deriving from its earlier Whig philosophy. Lubbock crossed sides in 1886 over the question of Irish home rule and became a Liberal Unionist. Lubbock mediated in both the London dock strike of 1889 and the coal porters' strike of 1890. Patton, *Science, Politics and Business*, 196–198.

22. Patton, *Science, Politics and Business*, 149, 150–151.

23. The Lubbock family home was not far from Darwin's home, Down House, and the families became close friends. In answer to the question of why he was a liberal, Lubbock later wrote, "in any complex community, the circumstances are continually changing, discoveries are made, improvements are suggested, and we cannot hope to maintain our population in comfort and prosperity, still less to diminish the suffering and misery around us, unless we are prepared to avail ourselves of all improvements and adapt ourselves to new circumstances and requirements." Cited in Patton, *Science, Politics and Business*, 165.

24. Roy M. MacLeod, "The Alkali Acts Administration, 1863–84: The Emergence of the Civil Scientist," *Victorian Studies* 9, no. 2 (1965): 85–112.

25. This could cause tensions in some individuals, for example J. D. Hooker, director of Kew Gardens. Hooker was dependent on paid employment, but his

relationships with his civil service superiors were fraught, and turned above all on questions of proper behavior; see Jim Endersby, *Imperial Nature: Joseph Hooker and the Practices of Victorian Science* (Chicago: University of Chicago Press, 2008), chap. 10.

26. Lord Rayleigh and F. J. Selby wrote of Glazebrook in his obituary: "He was not among the original great scientific thinkers to whom fundamental advances in knowledge and understanding of the universe are due, he had not so much the contemplative philosophic, creative, mind as the brain and quick grasp of the man of action, who is ever ready to turn his rapidly acquired and clearly apprehended knowledge to practical ends." *Obituary Notices of Fellows of the Royal Society* 2, no. 5 (1936): 53. Such qualities were particularly useful in scientific affairs in the metropolis.

27. Bernard H. Becker, *Scientific London* (London: Henry S. King, 1874), 53–71. One reason for the lack of attention is that many studies focus on the arts and design side of Society of Arts activities, rather than the more scientific and technological. In 1908 the society was granted the prefix "Royal" for its name.

28. The main general histories of the society are Sir Henry Trueman Wood, *A History of the Royal Society of Arts* (London: John Murray, 1913), and Derek Hudson and Kenneth W. Luckhurst, *The Royal Society of Arts, 1764–1954* (London: John Murray, 1954). There are many articles in the *Journal of the Royal Society of Arts* on aspects of its history, notably by David G. C. Allan, but there is no overarching history of the society in the period under consideration here.

29. Examples include Sir John Wolfe-Barry (1836–1918), civil engineer and designer of one of London's best-known landmarks, Tower Bridge, who served on commissions on the Port of London and on Traffic in London. Another was the petroleum consultant Boverton Redwood (1846–1919), who served as technical advisor to the City of London and the Port Authority. See *Oxford DNB* for both.

30. This is by no means a complete list of Trueman Wood's circle, but has been drawn from those who served regularly on the council and who featured in his daughter's unpublished memoir; see Freda Fisher, "Sir Henry Trueman Wood, written by his daughter, Freda Fisher" (née Wood), c. 1928–1940, typescript, Library of the Royal Society of Arts.

31. Lyon Playfair (1818–1898) remained closely involved with chemical institutions and, unlike other scientific men, close to the royal family. He was the first scientist to receive a peerage, in 1892, proposed by Gladstone. Like others of the scientific elite, he also acquired commercial directorships later in his career. T. W. Reid, ed., *Memoirs and Correspondence of Lyon Playfair* (1899; repr., Jeminaville, Scotland: P. M. Pollak, 1976); *Oxford DNB*.

32. A. J. Meadows, *Science and Controversy: A Biography of Sir Norman Lockyer* (London: Macmillan, 1972); and *Oxford DNB*.

33. For example, the Physical Society was founded by Frederick Guthrie in 1873 at the Royal College of Science in the teeth of opposition from the Royal Society: see Graeme Gooday, "Teaching Telegraphy and Electrotechnics in the Physics Laboratory: William Ayrton and the Creation of an Academic Space for Electrical Engineering 1873–84," *History of Technology* 13 (1991): 73–114.

34. It is notable, for example, that none of the elites discussed here was involved in the development of Fabian socialism, or close to Beatrice and Sidney Webb's efforts to influence the LCC under the Progressive party. Nor were any scientists included among the members of the Coefficients dining club established by the Webbs in 1902.

35. "Frederick Bramwell (1818–1903)," *Oxford DNB*.

36. C. A. Russell, *Sir Edward Frankland: Chemistry, Controversy and Conspiracy in Victorian England* (Cambridge: Cambridge University Press, 1996), provides a detailed account of Frankland's career in London. Frankland in later years neglected his teaching in favor of his water analysis business.

37. M. H. Port, *Imperial London* (New Haven: Yale University Press, 1995), 269.

38. Martin Campbell-Kelly, "Data Processing and Technological Change: The Post Office Savings Bank, 1861–1930," *Technology and Culture* 38 (1998): 1–32. There may also have been some attempt to design post offices in accordance with Fordist principles (information from design historian Stephen Hayward).

39. Port, *Imperial London,* 136. Baedeker was always impressed by statistics of size, and Blythe House was no exception; see his *London and Its Environs*, 15th ed. (Leipzig: Karl Baedeker; London: Dulau & Co., 1908), 387. Ironically this building now houses the stores for the Victoria & Albert and Science Museums, but is threatened with redevelopment on account of its land value.

40. William Preece twice (1880 and 1893), Edward Graves (1888), and Sir John Gavey (1905). Frank Scudamore (1823–1884), Second Secretary of the Post Office, who set up the savings bank and masterminded the acquisition of telegraphs, was also an early president (1873).

41. "Sir William Henry Preece (1834–1913)," *Oxford DNB*. A partial list of his writings and lectures appears in the laudatory biography by E. C. Baker, *Sir William Preece, F.R.S., Victorian Engineer Extraordinary* (London: Hutchinson, 1976), 355–362.

42. Daunton's history excludes the telegraph and telephone, and thus makes no mention of Preece; see M. J. Daunton, *Royal Mail: The Post Office since 1840* (London: Athlone, 1985). Bruce Hunt is naturally more concerned with Heaviside,

who was both right scientifically and shabbily treated by Preece; see Bruce J. Hunt, *The Maxwellians* (Ithaca: Cornell University Press, 1991), 58–61, 160–174.

43. See the list in Baker, *Sir William Preece*. Preece's lectures were generally published in the electrical press, and he occurs frequently in the pages of the *Journal of the Society of Arts*. See also the references to Preece in Graeme Gooday, *Domesticating Electricity* (London: Pickering and Chatto, 2008).

44. Described by a young telegraph maintenance assistant, P. R. Mullins, who was present; see Baker, *Sir William Preece*, 266–267.

45. S. Hong, "Marconi and the Maxwellians: The Origins of Wireless Telegraphy Revisited," *Technology and Culture* 35 (1994): 734.

46. S. Hong, "Styles and Credit in Early Radio Engineering: Fleming and Marconi on the First Transatlantic Wireless Telegraph," *Annals of Science* 53, no. 5 (1996): 431–465.

47. Marconi's approach to development, treating it more as a handicraft than as laboratory scientific engineering, fitted well with Preece's emphasis on the electrical engineer as a practical man, and thus incurred disparagement from scientists such as Heaviside and Oliver Lodge. They were contemptuous of Preece and said that he did not understand the difference between inductance and Hertzian waves. Nonetheless Preece helped mediate a settlement between Lodge and Marconi in 1906 over patents; see S. Hong, "Marconi and the Maxwellians," 746–747.

48. Henry Frith and S. Rawson, *Coil and Current or the Triumphs of Electricity* (London: Ward, Lock, 1896), 67–77.

49. Baedeker, *London and Its Environs*, 95. The reference to bankers as suitable referees underlines the GPO's place in the City and City life.

50. This was in the GPO "South" building. *Electrical Handbook for London* (London: Institution of Electrical Engineers, 1906), 156–162.

51. The history of the IEE is told in Rollo Appleyard, *The History of the Institution of Electrical Engineers (1871–1931)* (London: Institution of Electrical Engineers, 1939), and W. J. Reader, *A History of the Institution of Electrical Engineers, 1871–1971* (London: Peter Peregrinus/IEE, 1987).

52. The debates were chiefly over the role of inductance in clear signaling. Heaviside had been a member of the IEE but was struck off the Register in 1886, along with Guthrie, for failing to pay their subscriptions for over two years. Preece, who could not compete with Heaviside's mathematics, always tried to deny him any credit for the success of inductance loading, but due recognition was later restored when Heaviside was elected an honorary member of the IEE in 1909.

53. Reader, *History of the Institution of Electrical Engineers*, chap. 6. Also James Gillespie, "Municipalism, Monopoly and Management: The Demise of 'Socialism in One County,' 1918–1933," in Saint, ed., *Politics and the People of London*, 112–116.

54. Colonel R. E. B. Crompton (1845–1940) is another example of the entrepreneurial kind of career during this period. Crompton had a military background, advised the government on electrical projects in India, and remained a consultant to the War Office. He also set up an electricity supply company in London, then a company manufacturing electrical appliances, as well as taking an active role in the IEE and sitting on the committee of the National Physical Laboratory. See *Oxford DNB*.

55. The *Handbook* was compiled with the help of four journals serving the industry: *Electrical Engineer*, *Electrical Review*, *Electrical Times*, and the *Electrician*. For other examples of similar maps of the metropolis, see Felix Barker and Peter Jackson, *The History of London in Maps* (London: Barrie & Jenkins, 1990), 148–149. London trade directories often published maps, for example *Garcke's Manual of Electrical Undertakings & Directory of Officials* (London: 1896).

56. *Electrical Handbook for London*, 50, 58.

57. Ibid., 207.

58. The architects, Henry Percy Adams (1865–1930) and Charles Holden (1875–1960), epitomize a combination of traditional good building with modern planning and imaginative style. Throughout, the facilities were the most up-to-date available. Even the *Architectural Review* was impressed, commenting that the overall effect was "one of force, almost barbaric in character," a comment intended to convey the building's strength and modernity; see *Architectural Review* (March 1911): 159.

59. Since there was much talk during the 1906 visit of brotherly feeling, it is relevant here to consider the Gays' work on fraternal association: Hannah Gay and John W. Gay, "Brothers in Science: Science and Fraternal Culture in Nineteenth-Century Britain," *History of Science* 35 (1997): 425–453. However, the assertion of international fraternity may arguably also be an expression of national pride, in that the members of this youthful profession could now stand in the same rank as their confrères in any part of the world. Similar sentiments were found among gas engineers: see John Garrard and Vivienne Parrott, "Craft, Professional and Middle-Class Identity: Solicitors and Gas Engineers c. 1850–1914," in Kidd and Nicholls, eds., *The Making of the British Middle Class?*, 162–163.

60. Lubbock's first attempt to do so had been in 1873–1874. The 1882 Act created a National Monuments Commission, with an inspector and a schedule of monuments of national importance. While Lubbock was particularly concerned with preserving archaeological monuments, he was at the same time waxing lyrical in

the House of Commons about electric lighting. Patton, *Science, Politics and Business*, 151–153.

61. The architect of the new metropolitan drainage system was Sir Joseph Bazalgette (1819–1891), chief engineer to the Metropolitan Board of Works from 1855 to 1889. It had a major impact on London's sanitary situation and helped reduce the death rate. Bill Luckin and G. Mooney, "Urban History and Historical Epidemiology: The Case of London, 1860–1920," *Urban History* 24 (1997): 37–54.

62. Jonathan Schneer argues: "imperialism was central to the city's character in 1900, apparent in its workplaces, its venues of entertainment, its physical geography, its very skyline." Schneer, *London 1900: The Imperial Metropolis* (New Haven: Yale University Press, 1999), 13. For a dissenting view, see Bernard Porter, *The Absent-Minded Imperialists: Empire, Society, and Culture in Britain* (Oxford: Oxford University Press, 2004), 148–150. For an architectural survey up to 1915, see Port, *Imperial London*.

63. Sophie Forgan and Graeme Gooday, "'A Fungoid Assemblage of Building': Diversity and Adversity in the Development of College Architecture and Scientific Education in Nineteenth-Century South Kensington," *History of Universities* 13 (1994): 176–182.

64. Thomas P. Hughes, *Networks of Power: Electrification in Western Society, 1880–1930* (Baltimore: Johns Hopkins University Press, 1983), chap. 9, "London: The Primacy of Politics."

65. A brief biography is found in A. Stuart Gray, *Edwardian Architecture: A Biographical Dictionary* (London: Duckworth, 1985), 279–281. Gray is one of the few architectural authors who includes electricity generating stations in his section on building types; see 74–75.

66. Charles Stanley Peach, "Notes on the Design and Construction of Buildings Connected with the Generation and Supply of Electricity Known as Central Stations," *RIBA Journal* (9 April 1904): 307.

67. Reproduced in frontispiece to Peach, "Notes on the Design and Construction of Buildings," 278, and in Gray, *Edwardian Architecture*, 280.

68. Peach was careful to cultivate electrical engineers, and seems to have been friends with leading lights in the industry such as Sebastian de Ferranti (1864–1930) and Colonel R. E. B. Crompton. Ferranti was present when Peach read his RIBA paper, and spoke afterward in the discussion in slightly neutral tones about its value as a record of practice up to that point in a rapidly changing industry. Peach, "Notes on the Design and Construction of Buildings," 308–311.

69. Hannah Gay, *The History of Imperial College London, 1907–2007: Higher Education and Research in Science, Technology and Medicine* (London: Imperial College, 2007), 26–34.

70. The new institutes were People's Palace, Mile End (1887); South London Polytechnic, Southwark (1890); City Polytechnic (1891); Battersea Polytechnic (1891); South West London Polytechnic (1891, became Chelsea Polytechnic in 1895); Goldsmiths Technical Institute, New Cross (1891); Borough Polytechnic Institute, Woolwich (1892); North London Polytechnic (1893); West Ham Technical Institute and Library (1896, became North East London Polytechnic); Northampton Institute, Clerkenwell (1896). For a critical account of the differing views surrounding the development of the City and Guilds of London Institute and government legislation, see B. P. Cronin, *Technology, Industrial Conflict and the Development of Technical Education in Nineteenth-Century England* (Aldershot: Ashgate, 2001), 195–232. For a study of London's town halls, see *London's Town Halls: The Architecture of Local Government from 1840 to the Present* (London: English Heritage, 1999).

71. "Edmund Alexander Parkes (1819–1876), Physician and Hygienist," *Oxford DNB*; a brief history of the museum is given by B. P. Bergman and S. A. St. J. Miller, "The Parkes Museum of Hygiene and the Sanitary Institute," *Journal of the Royal Society for the Promotion of Health* 123, no. 1 (March 2003): 55–61. It was founded the year after the 1875 Public Health Act, which provided for the appointment of a medical officer of health by every sanitary authoritiy, amid continuing debates over metropolitan sanitation.

72. Bergman and Miller, "The Parkes Museum," 57.

73. A report of Galton's course of lectures on house drainage and refuse was published in the *Scientific American Supplement* no. 421 (26 January 1884), 6717, http://onlinebooks.library.upenn.edu/webbin/gutbook/serial (Project Gutenberg EBook accessed 10 August 2006). See also reports in the *Nursing Record* (17 May 1888): 75; (21 February 1889): 125.

74. The museum's lack of popular appeal was not perhaps enhanced by the inclusion of interesting specimens such as "Rat-gnawed pipes, showing the indented marks of teeth, clearly tracing the damage done by rats" in the 1891 catalogue; quoted in Bergman and Miller, "The Parkes Museum," 60.

75. H. G. Wells, "The Parkes Museum: The Place to Spend a Happy Day," *Certain Personal Matters* (London: William Heinemann, 1897).

76. Reported in the *Times*, 3 May 1899, 12, issue 35819, col. B. The museum and Sanitary Institute and their aristocratic patrons appeared regularly in newspapers such as the *Times*, in the form of reports on annual dinners, notices of lectures, and letters on improving aspects of sanitary education.

77. For histories of the South Kensington museums, see Anthony Burton, *Vision and Accident: The Story of the Victoria and Albert Museum* (London: V&A Publications, 1999); David Follett, *The Rise of the Science Museum under Henry Lyons* (London:

Science Museum, 1978); Xerxes Mazda, "The Changing Role of History in the Policy and Collections of the Science Museum, 1857–1973," *Science Museum Papers in the History of Technology,* no. 3 (1996).

78. Adrian Desmond, *Huxley: Evolution's High Priest* (London: Michael Joseph, 1997), 149.

79. This is particularly apparent in Huxley's 1876 essay on biological apparatus, where he refers to a "very interesting" collection of historic apparatus but dismisses the early compound microscope as little more than a toy. He passed rapidly on to modern microscopes and methods of physiological investigation, and, never missing an opportunity to make a political point, ended with a reference to the plan of the University of Breslau's Institute of Vegetable Physiology, a research institute which was unparalleled in England. T. H. Huxley, "Biological Apparatus," in *Handbook to the Special Loan Collection of Scientific Apparatus 1876* (London: Committee of Council on Education, 1876), 321–326.

80. This conflict was particularly evident with some of the art curators; see Burton, *Vision and Accident,* 132–136.

81. Owners often forgot—or did not want—to collect their artifacts at the end of exhibitions which were held on the South Kensington site, and these therefore ended up in the museum. In 1890, beside the art collections proper derived from the ostensibly "good design" function, the museum contained collections of construction and building materials, ship models, educational equipment, patents, food, fish culture (in which fish were hatched out each year), and scientific apparatus, together with machinery and inventions.

82. Norman Lockyer had joined the civil service as secretary to the Devonshire Commission (1870–1875). Donnelly, together with his scientific friends, then managed to engineer Lockyer's secondment to the Department of Science and Art as organizer of the 1876 exhibition, and allowed him to develop his solar physics work.

83. The exhibition was reported in journals such as the *Illustrated London News, Gardeners' Chronicle,* and *Saturday Review,* as well as in newspapers such as the *Times,* and there were copious reports in *Nature* and the *Journal of the Society of Arts.*

84. *Catalogue of the Special Loan Collection of Scientific Apparatus at the South Kensington Museum* (London: Committee of Council on Education, 1876); the second edition was a daunting nine hundred pages long. See also *Handbook to the Special Loan Collection of Scientific Apparatus,* cited earlier.

85. Conferences and free evening lectures were held in connection with the special loan collection of scientific apparatus at South Kensington Museum in 1876. William Spottiswoode articulated the view of the utilitarian and inspirational value

of instruments with great elegance, and his talk was reprinted in full in *Nature* (18 May 1876).

86. The memorial was published in full in *Nature* 14 (20 July 1876): 257–259.

87. John Hewish, *Rooms near Chancery Lane: The Patent Office under the Commissioner, 1852–1883* (London: British Library, 2000), chap. 13.

88. For example, Norman Lockyer reported on the Paris collections in 1873 for the Devonshire Commission, and Sir Henry Roscoe, the chemist, made an official report in 1886 on the Arts et Métiers.

89. This criticism was evident in contemporary guidebooks: "It contains the most magnificent collections, but they are over-crowded and therefore in part ill-arranged. This is why a visit to 'the *omnium-gatherum* of South Kensington' (as someone has called it), so often results in weariness of the flesh, vexation of spirit, and confusion of mind." *Darlington's Handbook of London* (Llangollen and London: Simpkin, Marshall, Hamilton, Kent, 1897–1898), 178.

90. W. S. Jevons, "Use and Abuse of Museums," in *Methods of Social Reform and Other Papers* (London: Macmillan, 1883). What follows is taken from this essay, originally written in 1881–1882 for publication in the *Contemporary Review*.

91. Despite this criticism, the museum also had to take care to avoid accusations of allowing itself to be used for purely commercial advertising, as manufacturers were apt to send it their latest models in an effort to use the museum as an additional shop window.

92. H. G. Wells, *Experiment in Autobiography,* vol. 1 (London: Gollancz, 1936), 209. This animosity was partly explicable by Wells's unhappy time as a student there. He formed strong antipathies, and so his account needs to be treated with caution.

93. Burton, *Vision and Accident,* chap. 9; Hermione Hobhouse, *The Crystal Palace and the Great Exhibition, Art, Science and Productive Industry: A History of the Royal Commission for the Exhibition of 1851* (London: Athlone, 2002), chap. 7.

94. Some of the collections continued to languish in the old exhibition galleries surrounding the site of the former Horticultural Society Gardens. Dave Rooney, "The Events Which Led to the Building of the Science Museum Center Block, 1912–1951," *Science Museum Papers in the History of Technology*, no. 7 (1997).

95. For example, "The Struggle to Create a Science Museum" is used as a subhead in Hobhouse, *The Crystal Palace and the Great Exhibition,* 279.

96. For example, Lyon Playfair wrote a key letter to the *Times* in 1897 suggesting rehousing the collections, citing the inevitable contrasts with museums in Berlin, Vienna, and Paris, which helped galvanize support for new buildings; quoted in Reid, ed., *Memoirs and Correspondence of Lyon Playfair,* 454–455.

97. Miriam R. Levin, "The City as a Museum of Technology," *Journal of History and Technology* 10 (1993): 27–36.

98. A selection of these images may be seen in Kenneth Chew and Anthony Wilson, *Victorian Science and Engineering Portrayed in The Illustrated London News* (Stroud: Alan Sutton/Science Museum, 1993).

99. Chris Otter, *The Victorian Eye* (Chicago: University of Chicago Press, 2008), 244. Otter emphasizes the combination of pragmatism and imagination that characterized the technological, administrative, and morphological problems of lighting an area as old and complex as the City.

100. Jevons, "Use and Abuse of Museums"; Baedeker, *London and Its Environs*, 396.

101. "The ruins of some latter-day South Kensington!" H. G. Wells, *The Time Machine* (1895; repr., London: Penguin 2005), 65.

102. Discourse at the Royal Institution by H. G. Wells, reprinted in *Nature* (6 February 1902): 326. My thanks to Harriet Waterfield for this reference.

103. *Report of the Committee Appointed by the Board of Trade to Make Enquiries with Reference to the Participation of Great Britain in Great International Exhibitions: Minutes of Evidence* (London: HMSO, 1908), Cd. 3773, xlix, 57.

104. Bernhard Rieger, "Envisioning the Future: British and German Reactions to the Paris World Fair in 1900," in Daunton and Rieger, eds., *Meanings of Modernity*, 149.

105. J. R. Piggott, *Palace of the People: The Crystal Palace at Sydenham 1854–1936* (London: Hurst, 2004), 31–65. The first chairman of the Crystal Palace Company, Samuel Laing, was at the time also chairman of the London, Brighton and South Coast Railways. This sort of entrepreneurial connection was not unusual.

106. The Agricultural Hall in Islington was well patronized by royalty, and took the name Royal Agricultural Hall from 1885. Alexandra Palace in Muswell Hill consisted of the rebuilt 1862 exhibition building from South Kensington, but it burnt down shortly after opening in 1873, and though rebuilt, never managed seriously to challenge its southern rival at Sydenham.

107. The reasons included disputes with the Horticultural Society over access to their gardens, the difficulties of persuading countries to contribute (the French refused to in 1873), general costs, and competition from international expositions elsewhere; see Hobhouse, *The Crystal Palace and the Great Exhibition*, 162–170. The Society of Arts also alleged that foreign contributions were insignificant; see [Trueman Wood,] Secretary of the Society of Arts, "Memorandum of Exhibitions," October 1886, p. 3; Royal Society of Arts, London, Archives, pamphlet no. 18.

108. The clearest account of the site and its relation to railway development is in *The Survey of London: Southern Kensington to Earls Court* (London: Athlone for the London County Council, 1986), xlii, 327–333.

109. These are described in Charles Lowe, *Four National Exhibitions in London and Their Organiser* (London: T. Fisher Unwin, 1892), and John Glanfield, *Earls Court and Olympia: From Buffalo Bill to the 'Brits'* (Stroud: Sutton Publishing, 2003). In addition to the American exhibition, the others were Italian (1888), French (1890), and German (1891).

110. Imre Kiralfy was invited to organize a recreation of Venice, complete with canals and gondolas (1891–1893), and of Constantinople, with hundreds of dancing girls on moving platforms which glided over the flooded interior (1893–1894). Glanfield, *Earls Court and Olympia*, 28–40; see also "Imre Kiralfy," *Oxford DNB* for Kiralfy's career.

111. This was explicit not only in the name but in the layout, which centered around the Court of Honor, the white buildings, and references to the "midway *plaisaunce*." See "The Franco-British Exhibition at Shepherd's Bush: Interview with Mr. Imre Kiralfy," *Review of Reviews* 37 (January–June 1908): 454–457. However, White City had a relatively short life as an exhibition venue, becoming a football ground after World War I. Both White City and Wembley included large sports stadiums as part of their design, and the fourth Olympic Games were held at White City in 1908.

112. Spencer Compton Cavendish, Marquess of Hartington and eighth Duke of Devonshire (1833–1908), was leader of the Liberal Unionists and president of the British Empire League, founded in 1894 to promote trade and harmony between Britain and its colonies. One of its goals was to improve steam and postal communications with India. Like his father, Devonshire was keen to encourage scientific, technical, and higher education; see *Oxford DNB*, and Wilson, *The University of London*, for his involvement with the university.

113. The land for Wembley was acquired by the Metropolitan Railway in 1889, which opened a station there in 1894 to encourage the use of Wembley Park as a center for leisure by Londoners living in the northwest.

114. Different approaches to the study of exhibitions may be found in Rieger, "Envisioning the Future"; or, for a focus on empire, in Peter H. Hoffenberg, *An Empire on Display* (Berkeley: University of California Press, 2001), and Annie Coombes, "The Franco-British Exhibition: Packaging Empire in Edwardian Britain," in Jane Beckett and Deborah Cherry, eds., *The Edwardian Era* (Oxford: Phaidon Press, 1987).

115. "The Franco-British Exhibition at Shepherd's Bush," 456.

116. The London Chamber of Commerce was founded at a meeting at the Mansion House, and always had its offices in the City. Although identified in many

ways as a "City" organization, its remit did cover the whole of the metropolis, but it was naturally concerned first and foremost with the promotion of trade in London. It was, however, usual practice for civic promotion in exhibitions to be handed over to the Chamber of Commerce.

117. *London Municipal Notes* 10 (January–June 1910): 178.

118. In addition, Lyon Playfair, T. H. Huxley, and John Tyndall were all vice presidents, thus adding a general scientific seal of approval to the enterprise.

119. *Official Guide to the International Health Exhibition, 1884* (London: William Clowes and Sons, 1884), 38. See also Annmarie Adams, "The Healthy Victorian City: The Old London Street at the International Health Exhibition of 1884," in Z. Celik, D. Favros, and R. Ingersoll, eds., *Streets: Critical Perspectives on Public Space* (Berkeley: University of California Press, 1995), 203–212.

120. *The Health Exhibition Literature*, vol. 7, *Health in Relation to Civic Life* (London: William Clowes and Sons, 1884), which contains five handbooks also separately published. The handbooks in the collected volumes do not always correspond with their contents as advertised in the *Official Guide*, perhaps because of later additions, or omissions and mistakes by the printers and binders.

121. *The Health Exhibition Literature*, vol. 8, *Health in Relation to Civic Life* (London: William Clowes and Sons, 1884). Three volumes (7–9) were devoted to the theme of health and civic life, the first a "handbook," the second covering conferences held (in which the Parkes Museum also participated), and the third public lectures.

122. *Official Report of the Japan British Exhibition 1910 at the Great White City, Shepherd's Bush, London* (London: Unwin Brothers, 1911). This report was drawn up under Kiralfy's direction and contained sections by different authors.

123. See Dirk Schubert and Anthony Sutcliffe, "Kingsway-Aldwych, 1889–1935," *Planning Perspectives* 11 (1996): 116–144. This was a long and hotly debated improvement, but one designed to solve long-standing traffic problems, as well as to clear an unwholesome area of streets and create a suitably "imperial" vista.

124. *Official Report of the Japan British Exhibition 1910*, 451. Those "mean streets" included Holywell Street and its environs, which Lynda Nead argues were at the center of the immoral "Modern Babylon" that so taxed mid- and late-Victorian city moralists. Nead, *Victorian Babylon*, part 3, "Streets and Obscenity," 161–189.

125. Hoffenberg makes this point in his analysis of the "Machines-in-Motion" section of the exhibition. Hoffenberg, *An Empire on Display*, 167.

126. Lowe, *Four National Exhibitions*, 48, 74.

127. Paul Lafage, in F. G. Dumas, ed., *The Franco-British Exhibition: Illustrated Review 1908* (London: Chatto & Windus, 1908), 281–284. What follows is taken from this source.

128. Moreover, Lafage named the experts concerned, mentioning "Miguel and Colmet-Daage [who] watch with tireless vigilance over the quality of our drinking water," and "the anthropometric service instituted by M. Alphonse Bertillon," to cite just a couple; ibid., 282–283.

129. In the 1904 St. Louis exhibition, however, the LCC displayed the shield used for excavating the Blackwall Tunnel, plus exhibits on drainage, sanitary buildings, and street paving and maintenance, showing drawings and photographs from Westminster; *Report of His Majesty's Commissioners for the International Exhibition, St. Louis, 1904* (London: William Clowes & Son, n.d.).

130. See "Patrick Geddes," *Oxford DNB,* and Helen Meller, *Patrick Geddes: Social Evolutionist and City Planner* (London: Rouledge, 1990). At the behest of Raymond Unwin, Geddes created an exhibition on approaches to city planning for the 1910 international Town Planning Conference, and developed it as a peripatetic exhibition. It was shown the following year at Crosby Hall, Chelsea, by the Sociological Society; *City Survey Preparatory to Town Planning* (London: Sociological Society: Cities Committee, February 1911), pamphlet in author's possession. Geddes's book was not published until 1915 (*Cities in Evolution* [London: Williams and Norgate]), and the remainder of his life was spent working overseas. His later reputation owed much to Lewis Mumford, who asserted his importance as a pioneer city planner.

131. The NPL exhibit became a standard feature at exhibitions beginning in 1908. A similar function was arguably served by the Wellcome Chemical Research Labs, another regular exhibitor, whose carefully designed cases showing pyramids of decorous bottles of chemicals created a stable yet aesthetic vision of expert science in the service of man, a point which was carefully emphasized in their literature. *The Wellcome Chemical Research Laboratories: Exhibits at the Franco-British Exhibition, London, 1908* (copy British Library 7957.de.24).

132. Likewise, in the 1910 exhibition, the London horse-drawn tramcar in the LCC exhibit, "a type which in London will soon become obsolete," attracted "special interest" and was displayed alongside the newer electric tramcar. *Official Report of the Japan British Exhibition*, 450.

133. See Adams, "The Healthy Victorian City," for its role in 1884.

134. This was approvingly noted by several publications, except *Nature*, which stuffily ignored Old London.

135. *International Inventions Exhibition: Official Guide* (London: Clowes, 1885), lix. One tenant of Old London was the recently formed Century Guild (1882), an Arts

and Crafts society, which designed a music room for one of the houses, thus providing an example of the best in "modern taste" made according to old handicraft methods. The room is depicted in *The Builder* (15 August 1885): 216.

136. *Illustrated London News* (8 August 1885): 19.

137. William Morris and SPAB, *A School for Rational Builders*, exh. cat. (London: Heinz Gallery, 1982); Hermione Hobhouse, *London Survey'd: The Work of the Survey of London 1894–1994* (Swindon: RCHME, 1994), 2–5.

138. As argued by Martin J. Wiener, *English Culture and the Decline of the Industrial Spirit 1850–1980* (1981; repr., Harmondsworth: Penguin, 1985), 67–71.

139. *Pictorial Souvenir of the International Exhibition and Old Edinburgh*, 1886; Alan Kidd, "The Industrial City and Its Pre-industrial Past: The Manchester Royal Jubilee Exhibition of 1887," *Transactions of the Lancashire and Cheshire Antiquarian Society* 89 (1993): 54–73 (with thanks to Nicholas Oddy and Sam Alberti for these references); Dorothy Rowe, "Georg Simmel and the Berlin Trade Exhibition of 1896," *Urban History* 22, no. 2 (August 1995): 225. Old Paris was illustrated in the *Illustrated London News* reports of the 1900 exhibition. A parallel could be drawn with the way that exotic native villages functioned in expositions, in that differences of time and space were reduced to manageable proportions within a single exhibit.

140. The Health Exhibition published precise attendance figures for each week together with sales of all the guides, catalogs, musical programs, and so forth; *Health Exhibition Literature,* vol. 19.

141. Gerald Maxwell, "Great International Exhibitions," *Westminster Review* 170 (1908): 680.

142. It is likely that the organizers had been impressed by the illuminations in the Paris Electrical Exhibition of 1881, reported in *Nature* and illustrated in publications of the electrical industry. See Jim Bennett et al., *1900: The New Age: A Guide to the Exhibition* (Cambridge: Whipple Museum, 1993), 60–62.

143. Figures are taken from [Trueman Wood,] Secretary of the Society of Arts, "Memorandum of Exhibitions," October 1886, Royal Society of Arts, London, Archives, pamphlet no. 18.

144. *Illustrated London News* (8 August 1885): 144–145.

145. *Illustrated London News* (23 May 1908): 753.

146. For example, see the reports in the *Illustrated London News* 86 (1885): 456; 87 (1885): 139.

147. *Nature* (4 June 1885): 107.

148. On the careful campaigns to popularize electricity and their relationship to supply problems, see Graeme Gooday, "The Expert-Consumer Relationship in Domestic Electricity," in Aileen Fyfe and Bernard Lightman, *Science in the Marketplace* (Chicago: University of Chicago Press, 2007), 231–268; and Graeme Gooday, "Profit and Prophecy: Electricity in the Late-Victorian Periodical," in Geoffrey Cantor et al., eds., *Science in the Nineteenth-Century Periodical* (Cambridge: Cambridge University Press, 2004), 238–254.

149. Dumas, ed., *The Franco-British Exhibition*, 299–300.

150. See research by Luke McKernan on early London cinemas at http://londonfilm.bbk.ac.uk.

151. For example, in 1908, the French railways presented a film of moving scenery, in contrast to the static photographs shown by the Great Western Railway, but in the 1910 Japan-British Exhibition, the London and North-Western Railway showed a film that was twenty minutes long. Dumas, ed., *The Franco-British Exhibition*, 259; and *Official Report of the Japan British Exhibition*, 126.

152. For example, two huge chimneys of the Electric Supply Company were visible from the grounds of the White City, which Kiralfy intended to render less visible with a judicious coat of paint, especially after they had been noticed on a visit by the Queen and the Empress of Russia. "The Franco-British Exhibition at Shepherd's Bush," 457.

153. The emphasis on the imperial—there was even a Festival of Empire Exhibition in 1911—reached its apogee in the 1924–1925 British Empire Exhibition.

154. As argued by Nead, *Victorian Babylon*, 212–215.

155. Patton, *Science, Politics and Business*, chap. 17.

156. Preece was a former student of King's, where he had been enrolled with the intention of following a military career, despite his father's background as a Welsh Wesleyan Methodist. Among Preece's companions in resignation was his and Trueman Wood's mutual friend Owen Roberts, clerk to the Clothmakers Company. Baker, *Sir William Preece*, 30–31; F. J. C. Hearnshaw, *The Centenary History of King's College London: 1828–1928* (London: George G. Harrap, 1928), 393. Both the LCC and the Treasury withheld grants from King's until religious tests were dropped. Preece's adherence to tests may have been simply nostalgic or part of the Anglican backlash against broader moves toward nondenominational education.

157. Otter, *The Victorian Eye*, 251.

158. In the case of electrical power supply, it is arguable that differences in output were in fact deliberately encouraged by companies to avoid their being taken over by a neighboring competitor, since competitive free enterprise was characteristic of London utilities.

4 THE COUNTERREVOLUTION OF PROGRESS: A CIVIC CULTURE OF MODERNITY IN CHICAGO, 1880–1910

ROBERT H. KARGON

PART I   INTRODUCTION

Rapid social, demographic, and technical change defined the birth and growth of nineteenth-century Chicago. Transformation on such a remarkable scale is always unsettling and disturbing, often unstable, and sometimes dangerous. For Chicago to be livable and prosperous, change had to be tamed.

Just as Manchester was the most exciting city of the 1840s, Chicago was the most exhilarating of the second Industrial Revolution. From a population of a dozen families in 1832, it grew—miraculously, given the poor site—to over half a million souls in 1880. The decade that followed witnessed an astonishing doubling of that population to over a million, enabling Chicago to surpass Philadelphia as the second largest city of the United States. The growth was made possible by technical change: by railway engineers who made Chicago one of the nation's great railway hubs; by civil and sanitary engineers who built roads, reversed river flows, and created high buildings; and by inventors and entrepreneurs who provided employment, drawing the population from the countryside (figure 4.1). Eventually, wholly new technical specialties—for example, gas, fire-safety, and electrical engineers—would be created to meet the needs of the metropolis.

Chicago's location, chosen for commerce, created tremendous problems in a period of rapid population expansion. The site was marshy, with poor drainage, and sewerage drained through one outlet into Lake Michigan. Population growth and industrial expansion after the Civil War turned the city into a disease-ridden locale permeated by an almost unbearable stench. Pollutants filled the rivers. Such conditions eventually prompted the city

FIGURE 4.1
Chicago. Elevated railway, 1905. Library of Congress, Washington, D.C.

to construct the Sanitary and Ship Canal (1889), and to reverse permanently the flow of the Chicago River in 1900, sending sewage into the Des Plaines River south of the city instead of into Lake Michigan.[1] But despite the difficulties, this period saw Chicago's dramatic economic ascendancy. As the center of a vast railroad and shipping network, Chicago processed and marketed the resources of its hinterland and provided itself and its countryside with an expanded new array of manufactures.[2] As historian Christine Rosen has put it, "Chicagoans both disliked and could not do without dangerous and polluting manufacturing activity."[3] To manufacturing activity we must also add the dangerous and often mysterious technological changes that directly affected the daily lives of all Chicagoans.

Construction using new techniques of civil engineering was omnipresent, to the extent that walking in the street offered its own kind of danger. Grade-level railway crossings claimed hundreds of lives each year; at the time of the Exposition of 1893, there were still three thousand miles of railway track, most at grade level, within the city limits.[4] Falling debris and holes in the ground claimed numerous other victims. Fear of fire—a demon of crowded urban spaces—was exacerbated by the memory of the Great Fire of 1871. The fire marshal at the World's Columbian Exposition was always on heightened alert, and even so, almost seventy-five fires occurred at the fair before it closed, and ultimately fire destroyed its remnants.[5]

Neither coal gas lighting nor the 1878 arrival of the electrical arc lamp, more than 150 times brighter, alleviated this fear of fire. Both systems necessitated the burning of hydrocarbons in the open air, and in the period 1880 to 1893 gas and electrical interests were intent on stimulating the fears of their rivals. The historian Harold Platt notes that some Chicagoans opposed the "mysterious" electrical force; for example, there were landlords who forbade "man killing wires." The promoters of electricity, not be outdone, stressed the problems of asphyxiation and explosion attendant on gas fires.[6] Fueled by the disputes between the Gas Trust and the advocates of electrical power systems, confusion and mistrust were widespread.

Along with this phenomenal demographic and technical change came class conflict. Chicago was a locus classicus of the growth pangs of industrial capitalism, with vast and pervasive class divisions. Although the city was only one battleground in a nationwide class conflict, Chicago's rapid and spectacular changes made the clash of interests especially raw. Socialist Eugene V. Debs maintained that "the marshy metropolis by the lake may boast of … sweatshops, slums, dives, bloated men, bedraggled women, ghastly caricatures of their former selves, babies cradled in rags and filth, aged children, than which nothing could be more melancholy—all these and a thousand more, the fruit of our present social anarchy."[7]

Violent unrest marked the 1870s through the 1890s, notably the Chicago manifestations of the nationwide "Great Upheaval": the railway strike of 1877, the Haymarket bombing of 1886, and the Pullman strike of 1894. The 1877 strike, which some have described as America's version of the Paris Commune, provoked a dramatic response by the city's great merchants and manufacturers. Marshall Field, the department store founder (figure 4.2), and the Citizen's Association donated money to the police to purchase heavier weapons, and armories were built to support an enhanced National Guard presence.[8]

What was to be done to deal with the unrest, and who was to do it? America in the nineteenth century was not a place for strong, activist

FIGURE 4.2
Chicago. Portrait photograph of Marshall Field, n.d. Library of Congress, Bain Collection, Washington, D.C.

government. Writing at a time when the city of Chicago was still embryonic, Alexis de Tocqueville, in his *Democracy in America*, claimed:

[Americans] are forever forming associations. There are not only commercial and industrial associations in which all take part, but others of a thousand different types—religious, moral, serious, futile, very general and very limited, immensely large and very minute. Americans combine to give fetes, found seminaries, build churches, distribute books and send missionaries to the antipodes. Hospitals, prisons and schools take shape that way. Finally, if they want to proclaim a truth or propagate some feeling by the encouragement of a great example, they form an association.

Where government is weak and not trusted, the people take matters into their own hands. Tocqueville continues: "Where in France you would find the government or in England some territorial magnate, in the United States you are sure to find an association. ... As soon as several Americans have conceived a sentiment or an idea they want to produce before the world, they seek each other out, and when found, they unite."[9] Through such associations the urban elites of Chicago, men of wealth, power, and influence, purposefully set about to construct a consensus about what the city should be and how it should proceed to become it.

The churn of events spawned new civic associations with interlocking directorates such as the Citizen's Association, the Law and Order League, and, above all, the Commercial Club, founded in December 1877, with Marshall Field and George Pullman as charter members, and Levi Leiter, Field's partner, as president. At monthly meetings, often with invited speakers, the business elite could discuss and plan for a better, safer, more orderly Chicago. They looked to a combination of force, educational and social improvement, and technological solutions. In the first several years, the topics of discussion included "the military as protectors of property, local and national"; "our sewerage: defects and possible remedies"; "should not the commercial prosperity of great cities be attended by the cultivation of art, literature and science?"; "nuisances: smoke, steam whistles, and bad streets"; " bridges, sewers (again)"; "the need for a school of industrial training"; and "cheap and good" worker housing. Over the years the Commercial Club was instrumental in founding the Chicago Manual Training School; in donating Fort Sheridan to the U.S. Army (after the Haymarket riots) and the site of the Second Regiment Armory to the state of Illinois;[10] in pushing for the drainage canal; and in planning, promoting, and carrying out the World's Columbian Exposition of 1893.[11]

The greatest names of Chicago commerce and finance were among its members; they included Harlow Higinbotham, J. Ogden Armour, Martin Ryerson, John Crerar, George Pullman, Charles L. Hutchinson, Samuel Insull, Cyrus McCormick, A. A. Sprague, and Frederic A. Delano, among many others. Educational, cultural, and religious luminaries such as William Rainey Harper of the University of Chicago and architect Daniel Burnham were also invited to membership.

The Commercial Club's avowed purpose was "advancing by social intercourse ... the prosperity and growth of the city of Chicago." It was an insider's club that promoted technical, cultural, and civic projects, such as the drainage canal, technical education, street cleaning, playgrounds, improvements in public education, the World's Columbian Exposition, and, eventually, a new urban design known as the Chicago Plan of 1909. The bedrock of their approach was to palliate conditions that they saw as the tinder of social conflict, and to create a consensus around the notion of smooth, managed technical and social change, or "progress" directed by "science." Very early on, in 1880, the Commercial Club discussed the attention that ought to be paid to fostering science, literature, and art, and just months later held a conference on the establishment of museums, libraries, industrial schools, and hospitals. Over the next three decades, members of the extraordinarily influential club would establish, fund, and often lead scientific, cultural, and educational institutions to ensure the successful propagation of the civic ideal they espoused.[12]

These Commercial Club members, as civic leaders, believed that an essential condition for domestic order and economic efficiency was the construction of a new civic ideal, an urban culture that, above all, encouraged order, stability, and the promise of prosperity. This civic ideal held that technology and science—and the ideas of progress coded within them—benefited all classes, and remained above politics and partisanship. This consensus or civic ideology has been described as a "revolution in values," turning upon a new appreciation of science and technology.[13] Under the banners of science, technology, commerce, and industry, social wounds would be healed, order maintained, and progress (moral and material) ensured. But first Chicago's history of rapid and sometimes uncontrollable change had to be tamed, repackaged as an inevitable march into the future, and thereby made acceptable to the populace.

The coalition that aimed to build consensus in favor of change and innovation comprised the city's "best men," and at times its "best women" as well: the business elite, along with their allies among the clergy, important cultural and artistic figures, and educational leaders. At the core of the coalition was great wealth.[14]

This chapter will focus upon three significant examples of the civic ideal made real, showing how powerful urban elites forged consensus regarding scientific and technical change. The three exemplars are:

1. Higher technical education in the form of the Armour Institute of Technology, formally opened in 1893, which soon became the premier producer of the "technostructure" of Chicago: the engineers, architects, and technologists of many kinds, who were responsible for planning, building, and maintaining the city's physical infrastructure.
2. The Field Columbian Museum, inaugurated in June 1894, which was originally conceived as much more than a natural history museum. Beyond the usual natural history categories, the Field Columbian included industrial arts, archaeology, anthropology, and other disciplines that would display American progress and skill, as well as the orderly succession of science and technology from past to present that was seen to be displayed at the World's Columbian Exposition.
3. The World's Columbian Exposition of 1893 itself, and growing out of the exposition, Daniel Burnham's Chicago Plan of 1909, an urban design that incorporated the civic ideal in a grand plan for the future of the city, a capstone to the vision of the past, a plan to be realized through the efforts of the alliance between business, the professions, and the arts.

The evidence clearly indicates that all three institutional exemplars were closely tied together through their leadership—their founders, authors, and directors, all of whom were members of Chicago's commercial, manufacturing, and financial elite.

PART II THE SHAPING OF A CIVIC CULTURE OF MODERNITY: THE ARMOUR INSTITUTE AND THE TECHNOSTRUCTURE OF CHICAGO

If progress—especially in industry—was to serve the economic and social roles envisioned for it, the creation of an educated, trained population was a necessity. Technical and technological education was a priority in producing human resources for a new industrial society. One of the early successes of the Commercial Club was the creation of the Chicago Manual Training School. In 1882 the Commercial Club raised funds for the establishment of this school; regular classes began in its new building at Michigan Avenue and Twelfth Street in 1884. It provided high school training in the "use of tools," with mathematics, drawing, and English included as part of the course. Its board of trustees included Marshall Field and George Pullman, and its faculty included a Ph.D. (H. W. Eaton) who taught physics, and a teacher seconded from the U.S. Navy (H. W. Parks), who taught mechanics and the design and construction of engines.[15] The Manual Training School was established squarely in the tradition of "self-help" philanthropy. The public aim, to give young men a leg up onto the ladder of self-advancement, also was directed at the perceived need of Chicago's industries for a better-trained, more disciplined workforce.

As the nineteenth century drew to its end, however, the need for more highly trained technical men for management, planning, and quality control became increasingly apparent. Schools of technology began to emerge with a new mission: to provide expertise and enterprise to manage the growing, increasingly technologically sophisticated needs of city and industry.

The institution of higher education that met that need was not the prestigious, recently founded University of Chicago, but another that began at about the same time: the Armour Institute of Technology. Philip D. Armour, the head of the world's largest meatpacking firm, established the Armour Mission in 1886; it offered classes mainly in religion and literature. Shortly thereafter, the Mission hired graduates of the Manual Training School to teach classes in woodcarving, mechanical drawing, and design. A turning point came in 1890 when Armour heard a sermon by the prominent minister Frank W. Gunsaulus calling for a million dollars to establish a school for young people to "help themselves." Armour donated the money, conditional only on Gunsaulus's devoting five years to the project. Influenced by the success of his friend Quinton Hogg's Regent Street Polytechnic, Armour worked with Gunsaulus to create the Armour Institute after the fashion of the Cooper Union of New York, the Pratt Institute of Brooklyn, and the Drexel Institute of Philadelphia. The coeducational Armour Institute's plan included a technical college (engineering, librarianship, and architecture), a scientific academy (secondary school), and programs in domestic arts, kindergarten, normal school, music, and commerce. The Institute emerged as nonsectarian, and was emphatically not a trade school like the Manual Training School.[16]

Between the initial plan for the school and its opening in 1893, the Armour Institute underwent a considerable evolution. Gunsaulus made a tour of technical schools in Europe, including one of the most influential of its day, the Technische Hochschule of Charlottenburg, Germany. Assessing the needs and opportunities of Chicago, Gunsaulus began to see the Institute's future in a new way. The new institution could, he foresaw, participate in the technological revolution then well under way by producing experts and expertise. Impressed by the cutting-edge technologies of electricity and telephony, bolstered by new methods of construction and by the increased demand for trained civil and mechanical engineers, Gunsaulus envisioned the Armour Institute as the Chicago center for a "professional career in engineering" at a time when Chicago Edison was wiring homes and public places, Western Electric was expanding its works, and daily life was being transformed by the science-related industries of electricity, heavy chemicals, pharmaceuticals, and steel.[17]

The Armour Institute opened officially on 4 September 1893. It had prepared itself well. The school received 1,600 applications for 300 places, subsequently expanded to 600. Electricity was the high-technology wonder of the 1893 Columbian Exposition, and the institute capitalized on the latest equipment demonstrated there. Wilber M. Stine, professor of electricity and electrical engineering, purchased for the Institute the lion's share of the best on offer there. One commentator reported, "Armour Institute purchase tags are tied on almost everything worth having and at the same time applicable to the equipment of an electrical laboratory." Stine, a member of the exposition's awards jury in electricity, boasted that the Armour Institute would have "the leading laboratory of electrical test instruments in this part of the world."[18]

In 1895 another stage in the Armour Institute's evolution was reached. The board of trustees, led by Gunsaulus and Philip D. Armour's sons, Philip Armour Jr. and J. Ogden Armour, changed the name of the school to the Armour Institute of Technology (AIT). The coeducational departments were moved to another site across town, and the new AIT would concentrate on training young men for careers in engineering, vastly enriching the technostructure of the city and enhancing its national reputation. The Armour Yearbook for 1895 expressed the desire of the school "to reach all classes" and "to educate the head, the hand and the heart," the last a nod to the Institute's roots as "avowedly a Christian school." Its departments included electricity and electrical engineering, mechanical engineering, domestic arts, library science, art, kindergarten teaching, and commerce, but courses were also taught in mining engineering, metallurgy, chemistry, and mathematics.[19]

In addition to Wilber Stine, the early faculty included Victor Alderson, mathematics; James Foye, Ph.D., chemistry; Ernest Cooke, mechanical and steam engineering; and Herman Haupt, M.D., Ph.D., mining engineering and metallurgy. Soon they were joined by Alfred Phillips, civil engineering; William T. McClement, chemical engineering; H. M. Raymond, mechanical engineering; Clarence Freeman, electrical engineering; and Lee De Forest, Ph.D., electricity. Alderson served as dean, moving on in 1903 to become president of the Colorado School of Mines. He was succeeded by H. M. Raymond.[20]

Electrical engineering was by far the biggest attraction at AIT. The department grew in a decade from two professors to a total of six. According to Stine's description, the curriculum emphasized "theory applied to electrical machinery, apparatus, practice, telegraphy, telephony, batteries, lamps, transmission and [power] station operation."[21]

The first class of twenty-one was graduated in 1897, with sixteen electrical engineers and five mechanical engineers. Degrees were added in civil engineering (1899), chemical engineering (1901), and fire protection engineering (1903). The Department of Fire Protection Engineering, unique in America, was headed by an AIT graduate, Fitzhugh Taylor (B.S. in E.E., 1900) who had been working at the Underwriters Laboratory. Starting in 1898 AIT engineering graduates were being hired by insurance companies as technical experts.[22]

In terms of numbers, electrical engineering graduates in the period 1897 to 1915 far outnumbered other specialties, followed by civil and mechanical engineering.[23] In 1904 the Armour Institute of Technology began to offer graduate courses in physics, mathematics, and electrical, civil, and chemical engineering. Early master's theses include the chemistry of construction materials and lubricating oils (chemical engineering), bridge, street, and sewer construction (civil engineering), and the design of electrical motors and transformers (electrical engineering).[24]

It is clear that the Armour Institute of Technology contributed mightily to the construction of a highly trained technical workforce for the city of Chicago. Over half the graduates of its first decade were working in the city in 1906. Of these, over a third were electrical engineers. AIT engineers worked in city inspection positions, in Chicago's industries, its public utilities, its educational institutions, and its insurance companies. AIT's semicentennial directory indicates that as of the 1930s a major proportion worked for public agencies.[25]

The dean of the School of Engineering, Victor Alderson, summed up Armour's first decade: "The day of the Untrained Man is past; the day of the technically-trained man is here. ... The incessant demand ... is for young men in the vigor of manhood, whose eyes, ears and hands, as well as minds, are trained to do the work demanded in modern industrial pursuits."[26] Morris Trumbull, in the *Armour Engineering Journal,* reinforced his dean's view: "the day has come wherein capitalists require that technically trained men take charge of heavy construction work and intricate machinery. ... At all times the fact must be appreciated that a technical education inculcates the method of attack in solving scientific questions rather than providing the solution itself."[27]

In providing the human resources for the technical sides of the public and private sectors in Chicago, the Armour Institute of Technology found itself in a position of local leadership. The University of Chicago, founded in 1892 on land in the Hyde Park section donated by Marshall Field, specifically rejected an engineering school. It remained for the Armour Institute to assume an advanced position in accommodating business, industry, and the public to the necessity, desirability, and inevitability of scientific and technical progress.

PART III   CREATING A USABLE PAST: THE ORIGINS OF THE FIELD COLUMBIAN MUSEUM

The World's Columbian Exposition of 1893 was an exemplar of Chicago's new importance as a rich and rapidly growing American metropolis. The Chicago Company, led by local businessmen, developed the fair. Its president was the Commercial Club's Harlow Higinbotham; most of its directors and major stockholders were likewise members of the club. The principal planner and director of works was architect Daniel Burnham, who would also become a club member.[28] Ostensibly mounted to commemorate the four hundredth anniversary of Christopher Columbus's voyage to the New World, the exposition was a celebration of America's industrial might, technical ingenuity, and wealth, as well as an open proclamation of Chicago's ascent to leadership. Chicago thereby announced that it was a great city and that, in its future, it would be an even greater one.

On 1 May 1893, President Grover Cleveland inaugurated the exposition by pressing a gilded telegraph key, which set in motion an electrical current that in turn gave life to engines, machines, fountains, whistles, and lights. What some called the greatest world's fair ever was open for business, for pleasure, and for edification.[29] Despite a severe national economic depression, in the six months that the exposition remained open, the proprietors recorded over 21.4 million paid admissions, the equivalent of approximately a third of the United States' estimated population at that time. Across a broad cross section of American opinion, the Columbian Exposition commanded serious attention: praise, condemnation, analysis, and puffery. It was generally agreed that Something Important had taken place.

The exposition was situated on the lake in Jackson Park, south of the center of the city, and was divided into three main sections. First, the "Court of Honor," or main part of the White City (as the fairgrounds was nicknamed), was at the south end of the site. Here, around the water basin, were arrayed the gleaming buildings devoted to Electricity, Manufactures and Liberal Arts, Administration, and Agriculture (figure 4.3). To the north lay the Wooded Island, the Fine Arts Building, the lagoon, and buildings devoted to Transportation, Fisheries, Mines and Mining, Women, and the pavilions of foreign countries and U.S. states. The third section was the amusement area, called the Midway Plaisance, which reached westward from the stately main site. It included the famous Ferris wheel, but also exotica such as the Dahomeyan Village, which brought African natives to Chicago to "convince white fairgoers of their racial superiority," according to some historians, or perhaps to give strong emphasis to the march of progress of technical civilization.[30] Like the anthropological exhibitions, historical displays at the exposition were mainly employed to underline the long way Western civilization had come, marking the unrelenting march of civilization that would lead inexorably to a better future.

The organizers of the exposition wished to preserve this clear message of inevitable progress after the gates of the fair closed. Even before the exposition opened, there had been suggestions that its collections could somehow be brought together in a great museum that would make the spirit of the exposition permanent. The result eventually took form as the Field Columbian Museum, the character of which was shaped by its roots (figure 4.4).

The Field Columbian Museum (now the Field Museum of Natural History) was the product of three factors. The first was the concept of cultural philanthropy. Chicago, now a worthy rival of New York and

FIGURE 4.3
Chicago. View of the 1893 World's Columbian Exposition. Library of Congress, Washington, D.C.

Boston, required the cultural muscle of those places, such as that provided by museums. The business community had recognized the value of cultural institutions very early on. In 1880, the Commercial Club discussed the establishment of museums, libraries, industrial schools, and hospitals as part of an effort toward the "cultivation of art, literature and science." One of the key figures backing this approach was Edward E. Ayer, a very wealthy lumberman and manufacturer of railroad ties. Frank Lockwood, his biographer, reports that "the leading captains of industry of his day were his intimate friends. He was everywhere counted as one of them." He aspired, however, to a larger vision. Lockwood continues, "He desired rather to endow coming generations with the imperishable riches of art and learning. ... In the process of inspired getting and generous giving he came to be a widely educated man [who] walked constantly in the company of men of genius, specialists, thinkers, artists, musicians and poets."[31] Ayer amassed over the years an important collection of Native American paraphernalia, pictures, and printed matter. He was a serious devotee of American ethnography, and his magnificent collection formed a major part of the

FIGURE 4.4
Chicago. Field Columbian Museum, c. 1901. Library of Congress, Washington, D.C.

exhibition of the Department of Anthropology at the World's Columbian Exposition. Ayer was one of the critical sparks of the effort to build a true museum out of the collections amassed for the exposition. His recounting of the story provides some flavor of how things worked among the wealthy and powerful in 1890s Chicago:

At the various Chicago clubs I came into familiar association with the leading men of the city at the table and at card games, so I began on all occasions to urge the importance of our getting material for a museum at the close of the World's Fair. There were several others who thought as I did—among the principal ones being George M. Pullman, Norman Ream and James Ellsworth. These men endorsed ... my remarks. Of course Marshall Field was the richest man we had among us in those days, so during our fishing trips and on social occasions I began to talk to him ... about giving a million dollars to start with.[32]

According to Ayer, Field was a hard sell, but eventually he was worn down. He pledged his million dollars, to which were added the pledges of fellow

Commercial Club members Pullman ($100,000), Harlow N. Higinbotham ($100,000), and George Sturges ($50,000).

The second factor that helped shape the museum was the interest and support of museum professionals, educators, and scientists. One of the early advocates of a natural history museum for Chicago was Frederick W. Putnam, professor of anthropology at Harvard and curator of its Peabody Museum. In May 1890 Putnam wrote a letter to the editor of a Chicago newspaper advocating a museum as well as urging that the forthcoming fair host an "ethnographic exhibition of the past and present peoples of America and thus make an important contribution to science."[33] Despite the opposition of the *Chicago Tribune* ("Dried up specialists mistake the purposes of the Fair") and the *Chicago Globe*, Putnam was appointed head of the Department of Ethnography of the exposition in February 1891.[34] Later that year, on 28 November, Putnam was invited to speak before the Commercial Club. In an address titled "Columbus Memorial Museum," Putnam advocated the establishment of a "grand Museum of Natural History" in Chicago to serve "students and lovers of nature" who, he argued, deserved to "be given equal opportunities for observation and investigation in the natural sciences." Putnam's museum, as proposed, would include departments of geology, zoology, mineralogy, botany, and anthropology.[35] The Reverend Frank Gunsaulus, president of the Armour Institute of Technology, recalling the history of the Field Museum in a letter to director Frederick J. V. Skiff, saw the museum as the result of the widely acknowledged educational function of the exposition. The world's fair, he wrote, "educated the mind of millions of our citizens. ... It was more than a great popular university. [It showed] the attainment of man ... and the processes of human development." The remaining task was to make it endure. Reverend Gunsaulus saw the push for the establishment of the museum as coming from "all our educational forces in the community" whose "citizens met together ... and a man was looked for who could and would meet the situation with adequate resources." It is at this point that Gunsaulus's narrative intersects those of Ayer and Marshall Field.[36]

The third force in the founding of the Field Museum was the lobbying effort of the Chicago Company of the World's Columbian Exposition, especially of its board of directors and its executive officers. In July 1893, Commercial Club member and Company executive S. C. Eastman published a letter in the *Chicago Tribune* advocating the establishment of a museum based on the collections of the exposition, the construction of which was then under way. The following month the directors of the exposition appointed the fair's president, Harlow Higinbotham, along with J. W. Scott and George R. Davis, as an action group to galvanize public sentiment for such a museum. One hundred leading citizens were convened, various forms of organization discussed, and eventually a group organized to incorporate as an independent museum company. The incorporators included Ayer, who served as chairman of the finance committee of the preliminary group.[37] The group's interest was sparked by the success of the World's Columbian Exposition in bringing together historical, anthropological, scientific, and industrial collections for exhibition. The fair was widely extolled—in Gunsaulus's words—as "a great popular university," and it was also the great symbol of Chicago's power and success. Though the exposition had to close, it would be shameful to let its crowning achievements slip away. The new museum, in the eyes of the fair's leadership, would be a memorial to Chicago's great moment, and a way of keeping it alive. The three founding forces of the museum came together as the Field Columbian Museum, sited at the only permanent building of the exposition, the Fine Arts Building, and formally dedicated on 2 June 1894.

The museum was organized as an independent corporate entity whose board of trustees reflected its prehistory. Members of the Commercial Club dominated the board: the museum's officers included Edward Ayer, as president, and Martin A. Ryerson, as first vice president. Harlow Higinbotham of the Marshall Field Company, also a major donor, chaired the powerful executive committee. As the Museum's first director, the board chose Frederick J. V. Skiff, who was the exposition's chief of the department of mines and mining. Skiff's responsibilities at the museum included the organization, staffing, equipping, and installing of the museum exhibitions.[38]

President Ayer convened the opening-day proceedings and Rev. Gunsaulus gave the invocation. Director Skiff presented an oration that included a detailed administrative history of the museum, but more importantly laid out the ideological position of the museum as seen by its proprietors. No previous period in history, Skiff claimed, had been "so alive to the demands of progressive humanity. The annals of centuries do not contain such evidences of a quickened higher culture and uplifting of educational forces as have been evoked within the past few years on the shores of the lake that sweeps this park. The Exposition left its uneffaceable impress on the social, moral and intellectual development of the world." This museum, he continued, was founded "to gather up the truths of the sciences and the triumphs of the industries and preserve them as a perpetual benefit to mankind." He concluded with a triumphant flourish: "We have builded [sic] in a few short months a great structure on the broad highway of progress. Science and industry have entered its portals hand in hand."[39]

The collections and exhibitions of the new museum reflected this philosophy—science and industry marching together toward the future—and the triple roots of the organization. They fulfilled the museum's goal of demonstrating the progress of the mind and the hand of mankind in the departments of natural history (botany, zoology, ornithology, and geology), anthropology, and industrial arts (transportation, railways, textiles, gems, and jewels). The goal of mounting and maintaining a fit memorial to the Columbian Exposition was fulfilled by the Department of Columbus Memorial Collections and by the collection of memorial statuary exhibited in the rotunda.[40]

Although the Field Columbian Museum was built by creating a broad popular constituency, some of the resulting partnerships were uneasy. The cultural philanthropists and the museum professionals coexisted warily with those who wished to memorialize the exposition. For the public, however, these divisions were not readily apparent. The *Chicago Tribune* exulted that the opening ceremonies were "all like a memory of the fair." The *New York Times* termed it both "a distinctive monument of the great exposition" and "one of the greatest of American museums."[41]

It was Edward Ayer's intention, and that of the major donors, to make the Field Columbian Museum into a worthy competitor of the museums of New York and Boston. Since Skiff, originally a professional journalist, had been engaged in fair and exposition exhibitions since 1889, he began to think of himself as a museum professional. In 1895 he published an essay, titled "The Uses of the Museum," that stressed the museum's role as a uniquely positioned extender of knowledge. Because the museum illustrates ideas, he wrote, it has a powerful advantage in the process of education: "The picture that the eye photographs upon the mind is sharp ... and durable. ... The thought, the idea conveyed to the brain by speech or by type is often vague, incoherent and fleeting."[42] This "sharp" and "durable" picture makes the museum a fit complement to the university.

Skiff elaborated on these themes a decade later in an address, "The Uses of Educational Museums," delivered at the convention of the National Educational Association, in Ocean Grove, New Jersey, on 3 July 1895. "The educational museum should be," he said, "at once a laboratory and study for the scholar and a resort for popular culture." The museum has distinctive educational advantages: "The object itself is more accurate than a description of it; the eye is more reliable than the ear. Scientific instruction cannot be perfectly imparted except by object lessons."[43]

For 1894–1895, the museum's board of trustees included Ayer; Charles Hutchinson, former president of the Commercial Club; Allison Armour, scion of the meatpacking family; Harlow Higinbotham; and Frank Gunsaulus, president of the Armour Institute of Technology. The board members were definitely hands-on administrators. As chairman of the executive committee, Higinbotham operated as chief executive officer, even ordering that all correspondence coming into and going out of the museum be "vised [sic] and copied" by the office of the director. It was pointed out to him that "scientific men are extremely sensitive" and would be offended by this practice, and so the order was not carried out. [44]

Clearly the emphasis of the museum was shifting early on toward scientific research and education. Aided by pressures on space and curatorial time, the industrial and the memorial side of the enterprise faded quickly. By 1896, the Department of Industrial Arts was abolished; gone too were collections of printing and graphic arts. The industrial exhibits inherited from the World's Columbian Exposition were dispersed or returned to the original donors. Scientific work and exhibitions eclipsed all else. Skiff recounted that by November 1903, "very little material of [industrial or trade] character ... was within the Museum building."[45] The transformation was nearly complete. In 1906 the museum was reorganized and renamed the Field Museum of Natural History. Its remaining departments were those of anthropology, botany, zoology, and geology; and its role as the purveyor of a usable past was replaced by the search for scientific and cultural eminence. The idea of scientific and industrial progress advancing hand in hand was so firmly embedded in Chicago's civic culture that it was no longer necessary to proclaim it.

PART IV   THE DREAM OF THE WHITE CITY: THE CIVIC IDEAL EMBODIED

The historian Henry Adams, the great-grandson and grandson of American presidents, felt compelled to visit the World's Columbian Exposition two times. For Adams, "Chicago was the first expression of American thought as a unity; one must start there." America had made a fateful decision, as Adams saw it, and this verdict was delivered with clarity at the White City. The choice was between the older, simpler industrial and agricultural system and the new "capitalistic, centralizing and mechanical" order. Adams's keen eye saw the emerging alliance: "All one's best citizens, reformers, churches, colleges, [and] educated classes had joined the banks to force submission to capitalism," with "all its necessary machinery." Even though capitalism "ruthlessly stamped out the life of the class into which Adams was born," it "created monopolies capable of controlling the new energies that America adored."[46]

The message of the World's Columbian Exposition, the "dream of the White City," was an encapsulation of a civic culture decades in the making. It was the result, as Henry Adams so brilliantly intuited, of a determined effort by commercial, industrial, financial, and professional elites to respond to what they viewed as both a crisis and an opportunity.

The architect Louis Sullivan remembered Chicago this way: "Its people had one dream in common: That their city should become the world's metropolis."[47] The path to that future was guided by the city's power elite, that group of influential businessmen whose words were, if not law, at least powerful directives. The grasp of the new power elite encompassed not only finance, commerce, manufacturing, transportation, and the distribution of goods, but also the cultural framework within which the new "capitalistic, centralizing and mechanical" order operated.

The fair as iconic White City offered contemporaries a glimpse of what a well-ordered future might be. As previous fairs had done, this world's fair celebrated progress flaunted through material goods, displayed the wonders of new and dramatic technologies, and gave wide latitude to the presentation of national pride. What was different about this exposition, and what had the greatest impact, was the utopian aspiration the White City called forth.

The World's Columbian Exposition was designed to promote this civic ideal, not surprisingly as Commercial Club officers and members dominated the exposition's corporation. As mentioned earlier, Harlow Higinbotham, a Commercial Club stalwart, was president of the exposition, and the board of directors was replete with club members. The chairman of the committee on buildings and grounds, Edward T. Jeffrey, appointed Daniel Burnham as director of works. According to Louis Sullivan, "Burnham and Jefferey [sic] loved each other dearly. The thought of one was the thought of both." Burnham, the brain behind the plan of the fair, fit well into the increasingly corporate civic culture of the city. Disapprovingly describing the mood of the day as one of "intense commercialism," with a tendency toward "bigness [and] organization," Sullivan believed Burnham was the leading Chicago architect who comprehended and embraced the significance of these developments. Disparagingly, Sullivan said that in this business climate Burnham "sensed the reciprocal working of his own mind." Sullivan, of course, was bitter that Burnham chose what Sullivan saw as a faded and outmoded classicism as the organizing architectural theme of the fair: "Here was to be a test of American culture, and here it failed." The White City, according to Sullivan, was a "white cloud" that cast a "white shadow."[48] For his part, Burnham made no secret of his belief that aesthetics and business were not enemies. In an address before the Commercial Club, Burnham proclaimed, "Beauty has always paid better than any other commodity, and always will."[49]

The White City's physical message was one of order and symmetry. Constructed in neoclassical style, the clean, clear, white buildings resonated with conventional ideas of beauty, sending a message of calm, safety, cleanliness, and rationality. The *New England Magazine* article "What a Great City Might Be—A Lesson from the White City" attests to the success of the intended message: "[W]hen one entered the gates of the White City, he felt that he was in the presence of a system of arrangements which had been carefully and studiously planned. The city was orderly and convenient. … The problem of the architect, the landscape gardener and the engineer had been thoroughly thought out before the gates were opened. The result was preeminently satisfying."[50]

Commending the "cleanliness and neatness" of the White City, the author noted that these virtues were the result of diligence and foresight. Moreover, the "sense of absolute safety … was delicious." All of these good things, the author noted, were the result of the determination of the exposition planners to recruit the "best" of Chicago: "The best were called upon to produce the best ... the best minds ... the best artists ... the executive talent of that wonderful city. We saw what an ideal city might be."[51] Over a decade later Charles Zeublin, in his book *A Decade of Civic Development*, asserted that "for the first time in American history a complete city ... was built as a unit on a single architectural scale. ... The White City was unique in being an epitome of the best we had done, and a prophecy of what we could do."[52]

The "best we had done" was not only aesthetic but practical. While the visual aspect of the White City spoke of beauty, the unavoidable message of the exhibitions within the buildings was that aesthetic, moral, and material progress rested upon science, technology, and corporate enterprise. The *Official Guide* declared that the past quarter-century had been "the golden age of American enterprise, American industry and American development. Wonders have been achieved in every branch of thought and in every line of trade."[53] Electricity was the cutting-edge technology of the day, and this quicksilver force was used to dramatic effect at the fair. Because electricity was still relatively new, scarce, and viewed as dangerous, exposition planners made special efforts to present it as safe, efficient, beneficent, clean, and nearly miraculous. The exposition's power plants supplied over three times the capacity of the entire city of Chicago, enough to power the ninety thousand lights that illuminated the grounds—a "Fairyland," enthused *The Dial*—along with startling novelties, including the signature giant Ferris wheel, the moving sidewalk, the elevated electrical railway, the huge fountains, automatic doors, and the electric automobile.[54] The Electricity Building featured the General Electric Company's eighty-five-foot-high "Tower of Light," with five thousand colored lights, accompanied by the waltz music of John Philip Sousa's band.[55] The same hall also contained a complete model kitchen that featured a dazzling glimpse of an all-electric future with fire alarms, carpet cleaners, dishwashers, ironing machines, stoves, hot plates, and washing machines.[56] An all-electric villa featured electric doors, elevators, burglar alarms, and cigar lighters.[57]

The World's Columbian Exposition was an early example—and perhaps the first—of what became a typical American trope: techno-nostalgia. Burnham was, aesthetically and philosophically, a classicist. The White City sought a pleasing harmony through order, symmetry of line, and objects of art. Burnham's goal was to influence not only the "highly educated part of the community," but also "the masses," though "the beauty of its arrangement and of its buildings."[58] In this he succeeded beyond all expectations. However, the exposition was also intended to dazzle with its technological modernity, For progress defines itself with and for technology. The serenity of the classical aesthetic cloaks the technological muscle beneath, while the comfort of the past makes the future palatable.

The exposition was destroyed in a disastrous fire the summer after it closed. The impact of the White City, however, was felt for years afterward. One of its unplanned outcomes was the emergence of Daniel Burnham as one of America's most famous city planners.

In 1901, Burnham was chosen, along with Frederick Law Olmsted Jr., to lead a commission to develop a new plan for Washington, D.C. Burnham and Olmsted selected Charles McKim and Augustus Saint-Gaudens to join the group. Senator McMillan of the Senate Committee on Washington, D.C., served ex officio. As his first extended city planning experience, the Washington project encouraged Burnham to concentrate his energies on the planning discipline. In his article "White City and Capital City," written in 1902, Burnham both draws a historical line from the 1893 exposition to the needs of America's cities at the turn of the century, and stresses the contemporary demand for planning:

[I]t is not so much money that is wanted to shape municipal improvements in response to the growing taste of the American people as it is a general, a well-thought-out plan—a plan that reaches out not merely through the life of one throw of the political dice, but beyond men and seasons and policies, for a century. What is logical is also beautiful. The monuments of Pericles, reared in the zenith of Attic supremacy, are logical. The pilgrims of twenty-four centuries say that they are also beautiful.[59]

Burnham's planning career skyrocketed. He was chosen by Cleveland's Progressive mayor, Tom Johnson, to head that city's Group Plan Commission in 1902. By 1904 he was working on plans for San Francisco and for Manila and Baguio, the capital and summer capital of America's new Pacific colony, the Philippines.[60] Meanwhile, both the Commercial Club and the younger Merchant's Club fitfully discussed inaugurating a Chicago plan. In 1903 the Commercial Club appointed its own member, Burnham, to a nine-man committee to formulate such a plan. Thus, when Frederic Delano and Charles Norton of the Merchant's Club approached Burnham to create a comprehensive plan under their club's sponsorship, at first he declined. In 1906, when convinced that the Commercial Club would offer no objections, Burnham finally relented and assumed directorship of a large project that would result in the *Plan of Chicago* of 1909.[61] From 1906 to 1909 Burnham and his assistant Edward Bennett led a team, the Plan Committee, in a vast effort funded by the Commercial Club (which by then had merged with its junior rival the Merchant's Club) to produce the plan. Hundreds of specialists—architects, transportation experts, and sanitary, railroad, electrical, and mechanical engineers, as well as other experts—met in over two hundred meetings. Burnham's staff of architects, artists, and draftsmen labored to incorporate their ideas into a meaningful whole (figure 4.5). The resulting document, the *Plan of Chicago Prepared under the Direction of the Commercial Club during the Years MCMVI, MCMVII and MCMVIII*, lists Burnham and Bennett as authors, and Charles Moore of the American Institute of Architects as editor.

The first chapter dives into the fray. Rapid urbanization and its attendant problems, the authors propose, have led cities such as Chicago to realize that "a plan for a well-ordered and convenient city is … indispensable." The premise of the plan is that an American city like Chicago "is a center of industry and traffic." Just as importantly, the city has a symbolic mission to fulfill, "a dignity to be maintained." The origins of the plan lay expressly in the exposition; the energy of the plan depends on what Burnham calls "the spirit of Chicago … impelling us to larger and better achievements for the public good."[62] The second chapter's historical review highlights the importance of the city of Paris as a model. Considerable space is devoted to Haussmann's achievement and to the illustration of Eugène Hénard's proposed plan, which utilized radial arteries and an inner circuitous boulevard to enhance the symbolic value of public monuments and administrative buildings. The chapter concludes with an extensive summary of Burnham's planning exploits in Washington, D.C., Cleveland, San Francisco, and Manila.[63] The heart of the *Plan of Chicago* is chapter seven, titled "The Heart of Chicago." This chapter details the plans for boulevards, streets, parks, and parkways; relocation of railway terminals and transshipment points; improvement of the lakefront; development of a suburban highway system; and most importantly the building of a Civic Center and

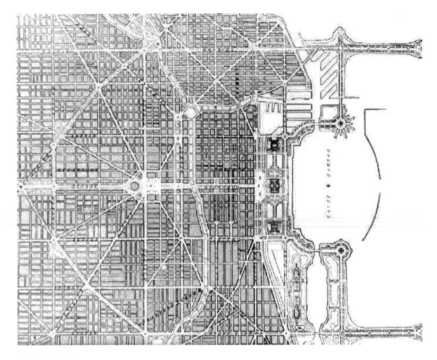

FIGURE 4.5
Chicago. "Plan of the Complete System of Street Circulation," from Daniel H. Burnham and Edward H. Bennett, *Plan of Chicago* (Chicago: Commercial Club, 1909). Photo courtesy of Special Collections, Kelvin Smith Library, Case Western Reserve University, Cleveland.

a center of intellectual life called Grant Park, all "so related as to give coherence and unity to the city."[64]

The plan had three main goals, growing out of its roots in the 1893 exposition. The first aim was the creation of the "City Inspirational," comprising the monumental Civic Center and the Grant Park intellectual center. At the heart of the latter is the future home of the Field Museum, described in the *Plan of Chicago* as "one of the important buildings of the city," meant to "gather under one roof the records of civilization culled from every portion of the globe, and representing man's struggle through the ages for advancement."[65] Along with the Field Museum was proposed the Crerar Library, the Art Institute, and other intellectual and aesthetic ornaments that would serve "those who are drawn by curiosity and those who come to study."⁶⁶ The crowning glory of Chicago, however, was be the Civic Center, composed of monumental public buildings, which would be to Chicago "what the Acropolis was to Athens or the Forum to Rome and what St. Mark's Square is to Venice—the very embodiment of civic life." Such a center would "typify the permanence of the city, record its history and express its aspirations."⁶⁷

The second goal was the "City Efficient." The plan proposed interlocking circulation systems of traffic, freight, passenger rail, and pedestrian movement. Within the city a new freight distribution system would be implemented, with freight terminals located under four passenger terminals. Southwest of the city would be a major freight-handling center, facilitating transfer of goods and the segregation of freight headed for other destinations from that destined for the city itself. The plan projected a regional highway system connecting within the city to a system of circuit and radial streets, like those of Hénard, and with a grand boulevard curving through the Civic Center.⁶⁸

A third goal was, of course, the "City Beautiful." Burnham's own vision was nothing less than lyrical:

Before us spreads a plantation of majestic trees, shadowing lawns and roadways upon the margin of the Lake. … We float by lawns where villas, swan-like, rest upon their terraces, and where white balustrades and wood-nymphs are just visible in the gloaming. The evening comes, with myriad colored lights, twinkling through air perfumed with water-lilies, and Nature enfolds us, like happy children.⁶⁹

The *Plan of Chicago* was published by the Commercial Club and appeared in 1909, beautifully printed and bound and lavishly illustrated with prints of paintings by Jules Guerin. It was apparent that the work of implementing the plan had only just begun. At a meeting of the Commercial Club in January 1908, Charles Norton suggested the formation of a permanent commission to oversee the dissemination of the plan. "The City Plan is a business proposition," wrote Norton, "and it should be developed under the direction and control of businessmen." Mayor Fred A. Busse appointed the Chicago Plan Commission and, at the Commercial Club's suggestion, appointed Charles Wacker, a former brewer, financier, and real estate magnate, as chairman. In 1911, Wacker was joined by Walter D. Moody as his assistant with the title of managing director.⁷⁰ To Moody fell the tasks of promotion and public relations.

Moody wrote and secured mass distribution of a 137-page primer on the plan that he titled *Wacker's Manual of the Plan of Chicago* (1911). His aim was to reach the schoolchildren of Chicago and their parents. He was able to persuade Chicago's superintendent of schools to adopt the first edition of the book officially. Over fifty thousand copies were printed.[71]

Moody was caught up in the scientism of the day. Historian Robert Wiebe describes in this period the rise of a "new middle class" as the bedrock of a professionalizing society. "Science" and "scientific" were the essential honorifics for this budding class.[72] For Moody, city planning and the promotion of city planning, when carried out systematically and methodically, were indeed "scientific."

Moody saw an acute need for thoughtful city planning. America was rapidly urbanizing, but the physical condition of the people in cities was deteriorating. Urban life, in Moody's opinion, had been debilitating, "more wearing upon the nerves than country life." The strain of city life had brought on insanity, debility, weakness, and shortened life. Something had to be done. The plan, in his view, promoted the efficiency, livability, and workability of the city in a scientific manner through a rational system of healthful parks, broad streets and boulevards, systems of transport, and civic pride. It was the last point—civic pride—that the monumental civic center addressed. Moody called such centers "vast civic temples"; their neoclassical design promoted a sense of stability, rationality, and symmetry that in turn would foster a sense of unity that Moody called "community virtue" or "community patriotism."[73]

Moody's reading of the *Plan of Chicago* owed much to his techno-nostalgia—a vision of the city that looked to the latest scientific and technical thinking in order to bring about the restoration of a more healthful and happier earlier way of life. In this, Moody's approach was similar to Burnham's own vision. In one of Burnham's last speeches before his death in 1912, he addressed the famous London Town Planning Conference in 1910, declaring that "a man who is accustomed to live in nature has a distinct advantage all his life over the purely town-bred man." In designing cities, Burnham said, planners must keep in mind that the citizen required parks and forests, for they would "balm his spirit needs. ... In the city of the future there should be no home not within reach of ... a public park ... where a town lies beside broad waters, keep all the shore for the people." The "air of the city of the future," Burnham predicted, "will be pure" because coal would be properly burned on-site and its wastes recycled. Energy would be transmitted to cities electrically, and would be used to run railroads. Well-paved streets would reduce dust. "The use of horses in a great city is near its end," he forecast, "because motor vehicles are becoming very cheap." The disappearance of the horse would end a "plague of barbarism." Congestion would be reduced in the city of the future through well-designed street systems, freight handling, and public transport—all key features of the plan. Burnham concluded with a general, stirring comment on the importance of making great, forward-thinking plans for a city:

But the question always arises when a given town is under consideration whether it would be wisest to limit suggestions to present available means, or, on the other hand, to work out and diagram whatever a sane imagination suggests. If the first be made your limit your work will be tame and ineffectual and will not arouse that enthusiasm without which nothing worth while is ever accomplished; it is doubtful, indeed, if even the meagre things proposed will be carried into effect. Such is humanity! You may expect support for a great cause, whereas men will yawn and slip quietly away from the obvious commonplace. ... Remember that a noble logical diagram once recorded will never die; long after we have gone it will be a living thing, asserting itself with ever-growing insistency.[74]

This long statement has come down to posterity in a more pithy formulation: "Make no little plans; they have no magic to stir men's blood and probably will themselves not be realized. Make big plans; aim high in hope and work, remembering that a noble, logical diagram once recorded will not die."[75]

Burnham's diagram was the White City itself.

The plan attracted its share of critics. Lewis Mumford, for instance, in *The City in History*, complained that the plan had "no regard for family housing, no sufficient conception of the ordering of business and industry themselves as a necessary part of any larger achievement of urban order."[76] While the *Plan of Chicago* was implemented only in parts, and even that partial success took decades to implement, its momentum continues to be at play.

PART V CONCLUSION

It has long been recognized, even before the term "hegemonic" became fashionable, that political and economic power are closely intertwined with culture and ideology. The evolution of Chicago's civic ideal is particularly instructive. In the 1880s, the British historian and constitutional lawyer James Bryce identified the real rulers of American cities as, on one hand, the industrial capitalists and, on the other, the "best men"—the cultural, educational, and social elite. The latter, Bryce wrote, were those who sought the best interests of the whole. The political scientist Moisei Ostrogorski, writing in 1902, saw both elements as part of a unitary capitalist elite that in his view ignored the general interest.[77] As Henry Adams perceived, the "best men" were swallowed up in the inexorable machine of the new finance capitalism, and its leaders were complex men who saw themselves as creators of a great good.

This unifying civic ideology captured both rulers and ruled. It was made necessary by the logic of technology, demographic growth, and the evolution of industrial capitalism. In 1848, Karl Marx and Friedrich Engels wrote an early analysis of the vibrant energy of industrial capitalism's "constantly revolutionizing the instruments of production, and thereby the relations of production and with them the whole relations of society." They described "all fixed, fast-frozen relations" as being "swept away, [and] all new formed ones become antiquated before they ossify. All that is solid melts into air, all that is holy is profaned."[78]

The rapid technological and social changes to which Marx and Engels referred at midcentury—"subjection of nature's forces to man, machinery, application of chemistry to industry and agriculture, steam navigation, railways, electric telegraphs, clearing of whole continents for cultivation, canalization of rivers, whole populations conjured out of the ground"[79]—could serve well as a description of the explosive history of the rise of the city of Chicago. It was just this situation—augmented by the new and even more rapid technical change of the 1880s and 1890s—to which the powerful bourgeoisie or business elite of Chicago so forcefully responded. The city's elite created institutions that would establish order and would manage and control technological and social progress. Through these institutions they built a vision of the future into the civic ideal.

The molders of Chicago's civic culture cunningly included the schoolchildren of the city in their target audience. The civic culture of modernity would be ensconced not only in the universities and technical institutes, in the city plans and appeals to voters, in the museums and art institutes, in the pulpits and newspapers and journals of opinion; it would reach down even to celebratory parades. In September 1913 in Palmer Park, the City of Chicago staged a civic pageant that included a tableau in which "Labor and Capital joined hands to symbolize their united commitment to harmony and progress." The pageant promoters described this joining of hands as enabling them "to free the city of the future" and symbolizing that "[t]he builders of the past pay tribute to the new city." The civic pageant was intended, as Percy MacKaye declared, to be an example of the "science of civic expression,"[80] or, as we have termed it, a civic culture of modernity.

NOTES

1. Libby Hill, *The Chicago River: A Natural and Unnatural History* (Chicago: Lake Claremont Press, 2000); Ann Keating, "The Sanitary and Ship Canal," *Encyclopedia of Chicago,* http://www.encyclopedia.chicagohistory.org/pages/1684.html. On public health and sanitation, see Harold Platt, *Shock Cities: The Environmental Transformation of Manchester and Chicago* (Chicago: University of Chicago Press, 2005).

2. William Cronon, *Nature's Metropolis: Chicago and the Great West* (New York: Norton, 1991), 263–309.

3. Christine Rosen, *Limits of Power* (New York: Cambridge University Press, 1988), 156.

4. W. T. Stead, *Chicago To-day; or, The Labour War in America* (London: Review of Reviews Press, 1894), 102.

5. Scott Knowles, "Inventing Safety: Fire, Technology and Trust in Modern America" (Ph.D. diss., Johns Hopkins University, 2003), 109.

6. Harold Platt, *The Electric City: Energy and the Growth of the Chicago Area, 1880–1930* (Chicago: University of Chicago Press, 1991), 28–30.

7. Eugene V. Debs, "What Is the Matter with Chicago?," *Chicago Socialist* (25 October 1902), http://www.marxists.org/archive/debs/works/1902/chicago.htm.

8. Carl Smith, *Urban Disorder and the Shape of Belief* (Chicago: University of Chicago Press, 1995), 109.

9. Alexis de Tocqueville, *Democracy in America*, ed. J. P. Mayer (New York: Harper, 1988), 513, 516.

10. Emett Dedmon, "A Short History of the Commercial Club," unpublished m.s., 1968, Box 1, f. 1, p. 11, Commercial Club Records, Chicago Historical Society.

11. John Glessner, *The Commercial Club of Chicago: Its Beginning and Something of Its Work* (Chicago: Privately printed, 1910), 13–14, 46, 131–136.

12. Ibid., 14, 189–198.

13. Robert Wiebe, *The Search for Order, 1877–1920* (New York: Hill and Wang, 1967), 147. Wiebe describes a "new middle class" as the bedrock of a "rising scientific-industrial society" (129).

14. W. T. Stead, British journalist and Christian socialist, reported on Chicago's mood in 1894 in the following way: "[Chicago men] have a trinity of their own of whom they think a great deal more than they do of the Father, Son and Holy Ghost. ... [Modern Chicago] subscribes with both hands to anything that is undersigned by the three Dii Majores of Prairie Avenue, Marshall Field, Philip D. Armour and George M. Pullman. ... They have the dollars and more of them than anyone else. Therefore, they of all men are worship-worthy." William T. Stead, *If Christ Came to Chicago!* (Chicago: Larid and Lee, 1894), 78. See also Thomas Schlereth, "Big Money and High Culture: The Commercial Club of Chicago and Charles L. Hutchinson," *Great Lakes Review* 3 (1976): 19.

15. John J. Flinn, *Chicago, the Marvelous City of the West: A History, an Encyclopedia and a Guide* (Chicago: Flinn and Sheppard, 1891), 229–230.

16. Robert H. Kargon and Scott G. Knowles, "Knowledge for Use: Science, Higher Learning and America's New Industrial Heartland," *Annals of Science* 59 (2002): 10–11.

17. James C. Peebles, "A History of the Armour Institute of Technology, 1896–1940," 1948, unpublished m.s., p. 8, Illinois Institute of Technology Archives; Kargon and Knowles, "Knowledge for Use," 12.

18. Peebles, "A History of the Armour Institute," 9, 11.

19. *Armour Yearbook for 1895*, 1–3. An extended account of AIT in the late 1890s can be found in Clifford Snowden, "The Armour Institute of Technology," *New England Magazine* 22 (1897): 354–372.

20. Peebles, "A History of the Armour Institute," 17–18, 28–33.

21. *Armour Yearbook for 1895*, 12.

22. *Bulletin of Armour Institute of Technology* (May 1908): 164–167; Peebles, "A History of the Armour Institute," 54.

23. *Bulletin of Armour Institute of Technology* (May 1916): 203.

24. "Register of Theses Written by Armour Institute of Technology Students," unpublished m.s., pp. 1–2, Illinois Institute of Technology Archives.

25. Illinois Institute of Technology, *Centennial Directory of Alumni* (Chicago: Clancy, 1939), 16; Kargon and Knowles, "Knowledge for Use," 14.

26. Victor Alderson, "The Economic Needs of a Technical Education," *Journal of the Western Society of Engineers* 7 (1902): 307–318.

27. Morris Trumbull, "The Technical Graduate," *Armour Journal of Engineering* (1903): 85.

28. Reid Badger, *The Great American Fair: The World's Columbian Exposition and American Culture* (Chicago: Nelson Hall, 1979), 51–55, 85–88.

29. Dennis Downey, *A Season of Renewal: The Columbian Exposition and Victorian America* (Westport, CT: Praeger, 2002), 3.

30. Robert Rydell, John Findling, and K. Pelle, *Fair America* (Washington, DC: Smithsonian Institution Press, 2000), 38.

31. Frank Lockwood, *The Life of Edward E. Ayer* (Chicago: McClurg, 1929), 76–77.

32. Ibid., 187–190. The donation amounts that follow are from the same source.

33. Donald Collier, "Chicago Comes of Age: The World's Columbian Exposition and the Birth of the Field Museum," *Bulletin of the Field Museum of Natural History* 40 (1960): 2–7 passim, 4.

34. *Chicago Tribune*, 16 September 1890, clippings file, Field Museum Archives, Chicago (henceforth cited as FMA).

35. Director's Correspondence, Box 4, Putnam files, FMA.

36. F. Gunsaulus to F. J. V. Skiff, 17 February 1917, Museum Foundation Files, Skiff file, FMA.

37. Museum Foundation Papers, Skiff file, 27, FMA; Lockwood, *Life of Edward E. Ayer*, 185–189.

38. "Skiff, Frederick James Volney," *Cyclopedia of American Biography* 12 (1904): 29.

39. Frederick J. V. Skiff, "An Historical and Descriptive Account of the Field Columbian Museum," *Field Columbian Museum, Publications* 1 (1894): 9–15.

40. Ibid., 6, 43–47.

41. *Chicago Tribune*, 3 June 1894, and *New York Times*, 4 August 1895, Historical Documents, Box 3, clippings file, FMA.

42. F. J. V. Skiff, "The Uses of the Museum," *Chicago Times Herald*, 29 April 1895, Historical Documents, Box 3, FMA.

43. F. J. V. Skiff, *The Uses of Educational Museums* (privately printed, 1905), 5–6.

44. "Annual Report of the Director," *Field Columbian Museum Reports* 1 (1894–1895): 3; Director's Correspondence, Box 7, Higinbotham file, FMA. Higinbotham to N. Ream, 5 February 1898; Owen Aldis to H. N. Higinbotham, 15 February 1898, Director's Correspondence, Box 7, Higinbotham file, FMA.

45. Museum Foundation File, 1916, FMA.

46. Henry Adams, *The Education of Henry Adams* (Boston: Houghton Mifflin, 1918), 343–345.

47. Louis Sullivan, *Autobiography of an Idea* (1924; repr., New York: Dover, 1956), 316.

48. Ibid., 314–319. Alan Trachtenberg, *The Incorporation of America: Culture and Society in the Gilded Age* (New York: Hill and Wang, 1982), chapter 7, describes the reaction to the White City.

49. Quoted in Paul Boyer, *Urban Masses and Moral Order in America, 1820–1920* (Cambridge: Harvard University Press, 1978), 264.

50. John Coleman Adams, "What a Great City Might Be—A Lesson from the White City," *New England Magazine* 14 (1896): 3–4.

51. Ibid., 7, 12–13.

52. Charles Zeublin, *A Decade of Civic Development* (Chicago, 1905), 59–60.

53. John J. Flinn, *Official Guide to the World's Columbian Exposition* (Chicago: Columbian Guide Company, 1893), 9.

54. Judith Adams, "The Promotion of New Technology through Fun and Spectacle: Electricity at the World's Columbian Exposition," *Journal of American Culture* 18 (1995): 49–50; John E. Findling, *Chicago's Great World's Fairs* (Manchester: Manchester University Press, 1994), 12; Downey, *Season of Renewal*, 89.

55. Downey, *Season of Renewal*, 92.

56. Thomas Schlereth, *Cultural History and Material Culture* (Ann Arbor: UMI Research Press, 1990), 282.

57. Julian Ralph, *Harper's Chicago and the World's Fair* (New York: Harper and Brothers, 1893), 196. See also Platt, *The Electric City*, 60–65.

58. Daniel Burnham, "A City of the Future under a Democratic Government," in *Transactions of the Town Planning Conference, Royal Institute of British Architects* (London: RIBA, 1911), 368–369.

59. Daniel Burnham, "White City and Capital City," *Century Magazine* 63 (1902): 619–620.

60. Mel Scott, *American City Planning since 1890* (Berkeley: University of California Press, 1969), 50–55, 60–65.

61. Kristen Schaffer, "Daniel H. Burnham: Urban Ideals and the *Plan of Chicago*" (Ph.D. diss., Cornell University, 1993), 269–279. See also her "Fabric of City Life: The Social Agenda in Burnham's Draft of the *Plan of Chicago*," in Daniel H. Burnham et al., *Plan of Chicago* (New York: Princeton Architectural Press, 1996), v–xvi.

62. Daniel H. Burnham and Edward H. Bennett, *Plan of Chicago Prepared under the Direction of the Commercial Club during the Years MCMVI, MCMVII and MCMVIII* (Chicago: Commercial Club, 1909), 1–8.

63. Ibid., 18–30.

64. Ibid., 121.

65. Ibid., 110, 114.

66. Ibid., 108.

67. Ibid., 117.

68. Nelson Lewis, *The Planning of the Modern City* (New York: John Wiley, 1916), 38–39; Jon Peterson, *The Birth of City Planning in the United States, 1840–1917* (Baltimore: Johns Hopkins University Press, 2003), 216–219; Gerald Danzer, "The Plan of Chicago by Daniel H. Burnham and Edward H. Bennett: Cartographic and Historical Perspectives," in David Buisseret, ed., *Envisioning the City: Six Studies in Urban Cartography* (Chicago: University of Chicago Press, 1998), 159–164. Indispensable is Carl Smith, *The Plan of Chicago: Daniel Burnham and the Remaking of the American City* (Chicago: University of Chicago Press, 2006).

69. Burnham and Bennett, *Plan of Chicago*, 110–111.

70. Michael McCarthy, "Chicago Businessmen and the Burnham Plan," *Journal of the Illinois State Historical Society* 63 (1970): 247–253.

71. Thomas Schlereth, "Burnham's *Plan* and Moody's *Manual*: City Planning as Progressive Reform," in Donald Krueckeberg, ed., *The American Planner*, 2nd ed. (New Brunswick, NJ: Center for Urban Policy Research, 1994), 138–140.

72. Wiebe, *The Search for Order*, 129, 147.

73. Quoted in Schlereth, "Burnham's *Plan* and Moody's *Manual*," 141–149.

74. Burnham, "A City of the Future," 372–378.

75. Charles Moore, *Daniel H. Burnham, Architect, Planner of Cities*, vol. 2 (Boston: Houghton Mifflin, 1925), 1921.

76. Lewis Mumford, *The City in History* (New York: Harcourt, Brace and World, 1961), 401.

77. David C. Hammack, "Problems in the Historical Study of Power in the Cities and Towns of the United States 1800–1960," *American Historical Review* 83 (1978): 323–349, discusses both Bryce and Ostrogorski among many other scholars.

78. Karl Marx and Friedrich Engels, *Basic Writings on Politics and Philosophy*, ed. Lewis Feuer (Garden City, NY: Doubleday, 1959), 10.

79. Ibid., 12.

80. Boyer, *Urban Masses and Moral Order*, 256–257; see also Clarence Rainwater, *The Play Movement in the United States* (Chicago: University of Chicago Press, 1922), 267.

5 "DAMNED ALWAYS TO ALTER, BUT NEVER TO BE": BERLIN'S CULTURE OF CHANGE AROUND 1900

MARTINA HESSLER

PART I   INTRODUCTION

Berlin's image has continually changed over time. In the eighteenth century, Berlin was perceived as a military city; at the beginning of the nineteenth century, Berlin was mocked as backward, mired in the Brandenburg soil. But by around 1900, Berlin had come to embody the modern city, thought of as a "space of experiment," or a "laboratory of modernity." And meanwhile, the German emperor was trying to strengthen Berlin's image as a Prussian city.

As many images as Berlin has had, and as often as they have changed, one factor has remained constant: the idea that Berlin is a fast-changing city. Karl Scheffler's famous argument—that Berlin was "dazu verdammt, immerfort zu werden und niemals zu sein" (damned always to alter, but never to be)[1]—neatly encapsulates the observation that Berlin is a culture of transformation. Without a doubt, Berlin is an outstanding site for analyzing a culture of change during the rise of the modern age. As Scheffler observed at the beginning of the twentieth century, Berlin was a place where a new "industrial culture" was pioneered. The city aimed at constructing this new culture with great "greed and passion."[2]

Berlin's population increased dramatically over the last quarter of the nineteenth century. In 1870, it had 800,000 inhabitants; by 1905, two million. The number of students rose in a comparably rapid way. In 1886, some 5,800 people were studying at universities in Berlin, out of a national total of about 30,000. By 1930, approximately 13,100 students were registered at Berlin's universities.[3] During this period, the Technical University (TU) and the Prussian Academy of Sciences were founded, along with numerous museums, libraries, technical colleges, and military training institutes.

Furthermore, in the last quarter of the nineteenth century, the economic and social structure of the city was substantially changed by the establishment of enterprises in the mechanical engineering, electrical, and chemical industries (AEG, Siemens, and Schering, for example). Institutions such as the Physikalisch-Technische Reichsanstalt (1887) and the Kaiser-Wilhelm-Institutes (1911) were founded, constituting a new type of applied science institute typical of the second industrial revolution. And Berlin became famous for its sewage system, copied in many other countries and based on scientific knowledge and the science-based concept of "public health." Urban infrastructures were reconfigured. Everyday lives changed as Berlin's citizens adopted modern, hygienic, science-based behaviors. Museums were built and public lectures given to spread scientific knowledge and demonstrate the value of science.

Within a very short time period, Berlin became an essentially *modern* city, in which science and technology would play a pivotal role in the economy, in politics, in creating urban infrastructures, and in everyday life. By the end of the 1920s Berlin had become an iconic modern metropolis and an international city. While Berlin is better known as a "modern" city in terms of its status in the 1920s as a center of the aesthetic movement of modernism, the first decades of Berlin's "modernity" were defined instead by science and technology.

PART II   A CULTURE OF CHANGE: CONCEPTS OF THE PAST, PRESENT, AND FUTURE

Scientists, industrialists, and local politicians built, as we will see, a dense network, pursuing together the project of transforming Berlin into a modern city. I will describe this culture of change by focusing on the actors concerned as well as on the question of how this tremendous change reconfigured time and urban space. I will discuss the ideologies and values of Berlin's elite, as well as the concepts that drove their reformulation of the past, construction of the present, and visions of a scientific-technical future.

In Berlin, the protagonists of change were particularly interested in constructing a new present. They aimed to create new scientific institutions, to build new infrastructure, to educate people about the value and benefits of science, and to popularize scientific progress. The past was not considered as constituting something of value in itself. As a city of permanent change, Berlin was (and still is) famous for neglecting its own past. Bertolt Brecht is said to have commented that Berlin, having no memory, was particularly ripe for revolution.[4]

That does not mean that a sense of the past did not play any role in Berlin's process of transformation into a modern city. Rather, we can observe a process of inventing traditions. Eric Hobsbawm has argued that many traditions which claim to be ancient are in fact recent in origin, sometimes invented during a single event or over a short time period.[5] According to Hobsbawm, traditions are invented more frequently at times of rapid social transformation. When Berlin underwent transformation around 1900, two new "traditions" were invented. While the emperor was attempting to invent a Prussian tradition, scientific elites reformulated the past in order to support the existence of linear scientific progress. Both "traditions" interpreted the past as providing continuity with the present, simultaneously opening a window onto the future. Berlin's museums and monuments exemplify well the specific situation of the city at the turn of the century, which was characterized by the tension between becoming the capital of a unified Germany and transforming into a modern city.

While modernity rejected the traditional values of the past and reformulated past events to correspond with the present and a predicted future, the idea of the future played a pivotal role. The idea of progress serves, as David Gugerli puts it, to create "certainty about an order of events in the future."[6] Narratives of continuity thus envisioned a modern future, which Berlin's scientific-technical elites believed would be technical, healthy, rational, and increasingly prosperous. That vision of a scientific-technological future would play a starring role in the industrial expositions.

PART III   BERLIN BEFORE THE "CULTURE OF CHANGE"

The speed and scope of Berlin's transformation is clear when one examines contemporary descriptions of the city before the 1880s. Around 1800 only seven thousand houses existed within the city walls, and the economy was dominated by craft and agriculture.[7] In 1865, an article in the *Berlinische Zeitung* complained about the provincial and traditional style of people's clothes, and criticized the city's architecture. The author argued that people's taste should be educated through special courses.[8] Even in the early 1880s, foreign visitors commented on how provincial and backward Berlin was. They described the unbearable hygienic conditions, dirty streets, darkness, and very bad smells.[9] Berlin was reputed to be one of the dirtiest and most evil-smelling of the European capitals.[10] These descriptions of Berlin as dirty, unhygienic, and dull were accompanied by the observation that it had a very backward population, who had migrated to Berlin from the countryside quite recently.[11]

However, another contemporary observer, Feodor von Zobeltitz, came to the conclusion that during the 1870s Berlin had become a *real* city. This, he claimed, was true not only because of the changing urban environment, but also as a result of the changing behavior of the city's people.[12] The socialist August Bebel stated that around 1870 Berlin left the phase of barbarism and entered the phase of civilization.[13] In speaking of "civilization," Bebel was describing a new urban lifestyle that was based on the scientific-technological transformation of the urban environment. Without a doubt, from the 1870s onward, the city began to lay aside its rather provincial image and was gradually transformed into—and perceived as—a modern city of science, industry, and culture.

PART IV   CONSTRUCTING THE PRESENT

In modern Berlin, a scientific-technical elite changed the urban environment, creating new urban spaces, new technologies, and new institutions. Some important scientific institutions and societies had existed as early as the eighteenth century, but the city became a widely recognized center of scientific research only in the last quarter of the nineteenth century.[14] Many scientists made Berlin famous: Robert Koch, Emil Dubois-Reymond, Hermann von Helmholtz, Max Planck, and, of course, Albert Einstein, to name but a few. Einstein even became a tourist attraction, as one travel guide from the early twentieth century reveals: many American women attended his public lectures, observing him through their field glasses.[15] Many Nobel Prize winners lived, researched, and taught in Berlin, in particular in the first decades of the twentieth century.[16] Rudolf Virchow played an especially important role in the city's transformation (figure 5.1), but other scientists, such as Robert Koch, were also involved.

FIGURE 5.1
Berlin. Portrait of Rudolf Virchow in the Institut für Pathologie. Berliner Medizinhistorisches Museum der Charité, Berlin.

Scientists played key roles in creating the culture of change. They built networks with local and national politicians and with industrialists, and they believed fervently in the benefits of science and technology. In contrast to Chicago, where the Commercial Club concentrated the main protagonists of change in one association (as Robert Kargon shows elsewhere in this volume), in Berlin change was conducted by various networks. Some personalities, such as Rudolf Virchow, were members of multiple networks, as we will see.

First, new science-based industrial concerns, and new types of research institutes devoted to industry and the economy, were established in the latter part of the nineteenth century. In 1879 the Technical University of Berlin was founded, integrating the Bauakademie and the Gewerbeakademie. Thus, around 1900, the rise of Berlin as a city of science and its rise as a modern industrial city were inseparably connected. Berlin's economy at this time was characterized by the existence of modern industries, which were above all science-based.[17] In 1871 two-thirds of Berlin's population worked in industry, while the other third worked in services and trade. The agricultural and crafts sectors, which had been the most important a few years before, had vanished.[18]

Recent research has produced much important work on the emergence of new institutions such as the Physikalisch-Technische Reichsanstalt and the Kaiser-Wilhelm-Institutes, and on the rise of science-based industries in Berlin.[19] Berlin was even called an "Electropolis," a term that became shorthand for the importance of Berlin's electrical industry in both the local and global economies.[20] David Cahan, in his work on the history of the Physikalisch-Technische Reichsanstalt, has shown how scientists and entrepreneurs worked together to found a new type of institute. He isolated and analyzed the scientific-industrial networks which led to the founding of the institute, commenting, for example, on a group including "Du Bois-Reymond, Hermann von Helmholtz, Heinrich Bertram, the Berlin school commissioner, Wilhelm Foerster, and Karl Schellbach," backed by the emperor and the crown prince.[21]

Scientific societies such as the Physikalische Gesellschaft and the Elektrotechnische Gesellschaft served to build ties between science and industry; a number of physicians and technicians were members of both societies.[22] Hubert Laitko has written about the density of these networks of scientists and industrialists, which often implied not only professional contacts but also friendship, and in the case of Werner von Siemens and Herman von Helmholtz, even close family links, since two of their children married each other.[23]

What is clear from this short sketch of the rise of new scientific institutes is that the process of Berlin's transformation into a modern, science-based city was driven by the close cooperation of scientists and entrepreneurs, who lobbied and fought for help from the state.

Secondly, a culture of change was enabled by the creation of new institutions aiming to educate the public. Scientists made great efforts to spread their findings. Science penetrated the public domain not only in the form of new technologies, such as electricity or public transport, or via the building of scientific institutions, but as a popular movement. Alexander von Humboldt's "Cosmos" lectures in Berlin had been famous in the early nineteenth century; his successors followed in the same tradition. The turn-of-the-century effort to popularize science, led by famous scientists, was intended to educate people and to spread a bourgeois culture that was closely connected with earlier movements for the popularization of science.[24] Scientists such as Rudolf Virchow, Robert Koch, and Hermann von Helmholtz gave public lectures. Three *Volkshochschulen* (adult education institutes) were opened between 1878 and 1902; libraries, where people could read and borrow books free of charge, proliferated as well. One of the most famous institutions of the popularization movement was Urania, the education center set up by Wilhelm Foerster and Wilhelm Meyer in 1888. It consisted of a "scientific theater," where scientific phenomena were explained to the public, as well as a laboratory where laypeople were allowed to do experiments on their own.

Despite these efforts, historian Hubert Laitko has come to the conclusion that the possibilities for laypeople to gain scientific knowledge remained modest.[25] Nonetheless, collections, museums, a zoological garden, and a botanical garden were also created. Besides serving the cause of scientific popularization, museums also helped to reformulate the past to correspond with the idea of linear, scientific-technological progress, as we will see shortly.

Finally, changing the present meant creating new technologies, new infrastructure, and new urban spaces. Around 1900 many new urban spaces emerged, their designs influenced by scientific considerations on the necessity of proper hygiene. Market halls were built in Berlin beginning in the

1860s. The central market hall at the Alexanderplatz was constructed in 1886—about thirty-six years after Paris's Les Halles.[26] Public toilets were installed, a central slaughterhouse was created, and parks and children's playgrounds—the modern commons—were constructed. New technologies such as electricity and public transport also changed urban space. A railway and streetcar system was built beginning in the 1870s, and in the early 1900s the Metro, or U-Bahn, was constructed. Simultaneously, industrial companies such as Siemens and Borsig moved to the periphery of Berlin, leading to the growth of commuting, the separation of private life from work, and thus to a new way of using space. Electricity began to be used for the lighting of public spaces, indicating the rise of a modern age and changing the way that public space was used. When official holidays such as the Sedanstag (commemorating the capture of Napoleon III at Sedan in 1870, effectively ending the Franco-Prussian War) were held, electrical displays symbolized the arrival of the modern technological age.[27] But one of the most important city-transforming infrastructural elements built at the end of the nineteenth century was the sewage system. This had been intensively discussed in Berlin for many years, before finally being constructed in the 1870s. Since Berlin's sewage system became a model for many other countries, it will be more closely examined here.[28] We will thereby have the opportunity to scrutinize in more detail how change in Berlin was driven by the cooperation of scientists, engineers, and politicians, all believers in the benefits of science and technology.

PART V BERLIN'S SEWAGE SYSTEM

From the mid-nineteenth century on, hygiene was a heated topic of debate in Europe. In the wake of cholera epidemics, ensuring the safe disposal of sewage water and a safe supply of drinking water became vital problems for European cities.[29] In Great Britain, Edwin Chadwick's "Report on the Sanitary Condition of the Labouring Population of Great Britain" was published in 1842, stressing the role of hygiene in the prevention of illness.[30] German cities were equally concerned with questions of urban hygiene and the possibility of improving public health, particularly after the International Hygiene Congress of 1852.[31] In Berlin at the beginning of the nineteenth century, sewage water from houses and streets flowed into gutters, which ran parallel to the houses and were mostly open, finally running into the river Spree. The system was a source of illness and bad smells and became the staple of jokes and sayings about Berlin's backwardness.[32] In 1872 Chadwick satirized Berlin's inhabitants, saying they could be recognized anywhere in the world because of the bad smell from their clothes.[33] In the same year, Anna von Helmholtz, wife of the scientist Hermann von Helmholtz, wrote to her mother, "The city administration will just think about the situation until a new pestilence occurs. ... Oh Haussmann! Would it only be possible to engage him for five years in Berlin!"[34]

The fast-increasing population constituted a problem for the city as early as the 1860s.[35] The scientific-technical elite was convinced that they could provide the best solutions to social and political problems. With their belief in the problem-solving capacities of science and technology, they aimed at changing traditional attitudes and habits in favor of a modern, hygienic, and rational lifestyle. It was clear to them that social and hygienic problems had to be solved technically in order to establish an objective given structure, which would ensure that people behaved in a modern manner. As Constantin Goschler has concluded, the construction of a sewage system could be interpreted as a technical reaction to increasing urban problems.[36] And as Horst Siewert has said, Berlin's elite was driven not by social utopian ideas, but by the desire to solve urban problems using scientific technologies.[37]

The construction of Berlin's sewage system is inseparably connected with the work of Eduard Wiebe, James Hobrecht, and Rudolf Virchow. Already in 1861 Berlin's *Baumeister* (master builder) Wiebe had developed a plan for a sewage system that would conduct the sewage water into Berlin's river.[38] Wiebe's plan had initiated a debate on alternatives for sewage removal. From this discussion finally resulted the so-called Hobrecht Plan, which was supported by Rudolf Virchow, as well as by Hobrecht's brother Albrecht, who became Berlin's mayor in 1872.

James Hobrecht is a central and contested figure in Berlin's history. Born in 1825 in Memel, East Prussia (now in Lithuania), Hobrecht is closely connected with Berlin's transformation into a modern city. Hobrecht did not finish his education, and became instead an apprentice surveyor. In 1847 he became a student at Berlin's Bauakademie. After finishing his study with the title of a Baumeister für Landbau in 1856, he took a second examination and became in addition a master builder for the construction of waterways and trains in 1858.

Klaus Strohmeyer has called Hobrecht the "prototype of a pioneer of modernization."[39] In 1859 the Königliche Polizeipräsidium assigned him the task of developing an urban plan for Berlin. Hobrecht's plan, which constitutes a very important part of the history of the city of Berlin, made its inventor famous. Its objective was to enlarge Berlin so that up to two million people could live in the city. Completed in 1862, the development plan has shaped the cityscape of Berlin up to the present day, creating Berlin's image as the biggest *Mietskasernenstadt* (city of tenements) in the world. During the early decades of the twentieth century, the plan was much criticized, since it led to high-density areas and thus intensified housing problems instead of solving them.[40]

While Hobrecht's work was negatively regarded by critics until the early 1980s, researchers since then have both reevaluated his work and emphasized his contribution to the new sewage system.[41] Hobrecht was one of the first German civil engineers who tried to combine urban planning with the construction of a sewage system, seeing them as inseparable.[42] In one of his main publications he described the necessity of creating the modern urban infrastructures which he aimed to establish in Berlin.[43] His concept for the sewage system differed from Wiebe's plan in that he did not support the idea of directing sewage water into the river. Hobrecht had traveled through Europe to examine different solutions, visiting Hamburg, Köln, and Paris, as well as London and other cities in England, at the beginning of the 1860s. As Wolfgang Ribbe comments, it is not quite clear how much Hobrecht's plan was influenced by his travels.[44] However, Hobrecht was certainly influenced by British civil engineer Baldwin Latham, as he said himself. And without a doubt, as Anna von Helmholtz indicates, Haussmann was a model present in people's minds whenever the sewage problem in Berlin was discussed.

Hobrecht's plan for a sewage system was sharply contested. Different social groups campaigned against it. Homeowners, for example, were afraid that they would have to pay the bill, while farmers complained that they would lose access to fertilizer if the plan was realized.

As already mentioned, Hobrecht was not the only agent promoting a new sewage system. The system was developed in close cooperation with the physician and politician Rudolf Virchow (1821–1902), who was almost the same age as Hobrecht. In 1867 a working group, consisting of members of Berlin's magistrate and its city council and including Hobrecht, was set up under Virchow's leadership. The working group's task was to create a scientific solution to the sewage question.[45] Virchow published a report in which he described the problems caused by the absence of a system. He pointed out that the presence of dangerous sewage from houses, stables, handicraft, and industry had been a topic of critical discussion in the city administration as far back as 1816, but nothing had been done.[46]

Virchow had become famous due to his theory of cell pathology. In 1839 he became a student at the military academy in Berlin, since his parents could not afford the university. After participating in the March revolution in 1848, he was fired from Berlin's university, where he had worked as a prosector. From 1849 to 1856 he was a professor of pathology in Würzburg, and then returned to Berlin to teach in that capacity at Berlin University. In addition to being a physician, Virchow was a politician. He was a member of Berlin's city council, the Prussian House of Representatives, and the Reichstag, and was a founding member and the president of Germany's Fortschrittspartei. Seeing himself as a social reformer, Virchow aimed to establish a "social medicine" based on scientific knowledge. Besides campaigning for a sewage system, he also worked on setting up public hospitals, parks, slaughterhouses, and so on. Without doubt, he was one of the most important personalities in Berlin's transformation into a modern city.[47]

As the main protagonists in establishing the new sewage system in Berlin, James Hobrecht and Rudolf Virchow were tied together by two common ideas: first, their strong concern with solving the social problems of fast-growing cities and creating healthy urban space; and second, the conviction that science and technology provided the fundamental bases for solving social and hygienic problems. As an engineer, Hobrecht was a prototypical modernist who thought in technical terms, while Virchow believed in science as the basis of public health.

The experience of these two men also helps us to see two interesting and converging factors. First, the scientific and technical elite voted for technical and scientific solutions for social problems. Second, the development of a new sewage system can be connected in general to the "rise of experts." Scientific expertise and scientific arguments played a crucial role in Berlin's transformation process, as is shown clearly by the example of the sewage system. Scientific arguments became decisive in the process of political decision-making, and scientific experts, headed by Rudolf Virchow, were assigned the task of finding solutions for the problem. However, we can also already observe the emergence of controversy associated with the use of experts. Different actors used different scientific arguments. The farmers who believed that the soil needed sewage water as fertilizer appealed to the scientific expert Justus Liebig in order to strengthen their arguments against the sewage system. Promoters of the sewage system, on the contrary, argued that it would be the only possible technically and economically reasonable solution.[48]

Virchow criticized the various interest groups, such as the homeowners, on the grounds that their arguments followed their own economic interests. Claiming that technical questions should not be decided by political groups or according to political arguments, he argued that there was only one reasonable technical solution, which had to be found by using technical and scientific criteria. Thus, questions such as those pertaining to sewage should be decided not by the political majority, but by scientists. Virchow emphasized the benefits of rational, scientific discourse and argued that scientists and engineers make decisions from an objective point of view. In the end, the city council followed the scientific and technical arguments of Virchow and trusted his scientific expertise: they decided on the Hobrecht Plan.[49]

Personal and technical cooperation, the authority of scientific arguments and particularly those of Virchow himself, as well as the political help of Hobrecht's brother, Albrecht, enabled them to drive through their solution, despite its being strongly contested. Both Virchow and Hobrecht argued that problems could not be managed simply by social welfare or social policy activities; rather, only "objective" technological solutions could solve the problems arising due to the increasing population of Berlin. During the last quarter of the nineteenth century, Hobrecht's and Virchow's activities changed Berlin's urban environment, the living conditions of its population, and the city's image.

PART VI   REFORMULATING THE PAST: A PRUSSIAN CITY AND A CITY OF SCIENTIFIC PROGRESS

As Paris and London developed into modern cities in the nineteenth century, at the same time they remained archives of their respective national histories. Urban elites in Berlin did not share such respect for maintaining the past; rather, they felt preservation was unnecessary, or even a hindrance to Berlin's transformation into a modern city. Indeed, in 1910 Karl Scheffler criticized Berlin's lack of respect for its history. Every generation, he felt, seemed to create itself anew. He fiercely attacked Berlin as being not German in character, but a city in the American style, a modern city with no history.[50] In fact, Berlin's rapid development led to frequent comparisons with Chicago.[51] Mark Twain, who stayed in Berlin for several weeks in 1891, was horrified: "The former Berlin has vanished. It seems to be vanished completely. No sign of it is visible anymore. Most quarters do not show any sign of earlier times. ... It is the newest city I have ever seen. Compared to Berlin, Chicago looks time-honored. Most of the city gives the impression of having been built one week before."[52]

These examples of Berlin's attitude toward its past are taken from the period of expansion and redevelopment that marked the last quarter of the nineteenth century.[53] After 1871 the government district was rebuilt, including the Reichstag and many new buildings for ministries. At the turn of the century the city council decided to demolish old quarters such as the Scheunenviertel and to rebuild them totally.[54] The narrow lanes, formerly home to cows and pigs, were destroyed, and with them the old way of life. As the medieval townscape vanished, new spaces emerged: Huge streets opened up for traffic, and in 1897 the first department store, Wertheim, opened its doors. In this urban modernity, old buildings and the rural lifestyle that had accompanied them were not regarded as a heritage worth preserving.

But the elites' concept of the past was a more complex matter. Berlin's elites were not completely indifferent toward the past. While they did not care for Berlin's urban heritage, they helped to build an enormous number of museums and monuments during the period under discussion. The museums and monuments that were built around 1900 expressed two different strategies: They served both to invent a past for Berlin—the new German capital—as an ancient Prussian city, and to legitimize science by presenting the history of Berlin—the modern city—as a narrative of linear progress. The tension between these two stories is typical of Berlin around 1900.

BERLIN AS A PRUSSIAN CITY

When Berlin became the German capital, it did not represent German history or the German nation. Because of Berlin's image as a modern—or even an American-style—city, the emperor built countless monuments in order to evoke tradition and a collective identity. Contemporaries talked derisively of the immense number of monuments built in this time—no less than 232 were put up by 1906. Wilhelm II stressed that these monuments served to extol the Hohenzollern dynasty and its achievements in bringing about German unity.[55] One of the most expensive and largest monuments built in the nineteenth century was the Siegessäule, or Victory Column.[56] Wilhelm I had originally decided to build the monument in 1864 (when he was King of Prussia), in order to celebrate Prussia's victory over Denmark. But after German unification he reconceived the Siegessäule as a celebration of Germany's unification and it became a national monument, celebrating the actions of the Prussian army as well as German unity. In its form the monument referred to classical antiquity. More importantly, the monument helped viewers understand the past by telling stories that constructed a linear history of Germany's nation-building process. The political message was clear: The military monarchy of Prussia was responsible for Germany's unification. The monument's bronze reliefs illustrated a chronological story, starting with the war between Denmark and Prussia in 1864 and ending with Germany's unification. Thus Emperors Wilhelm I and Wilhelm II underlined the Prussian roots of Germany as well as the military tradition of Germany in general and of Berlin in particular.

By displaying Prussian history, the German emperors aimed at providing a tradition for the newly built nation. In their historical narrative—which included monuments such as those to Bismarck (1901) and to Moltke (1904), as well as the so-called Siegesallee (Victory Avenue), where thirty-two statues of Prussian rulers were displayed—the past confirmed the existence of a linear progress which had its origins in Prussian history and which simultaneously hinted that Germany's future would be a continuation of the Prussian state. By reformulating the past in accordance with contemporary developments, such monuments reconstructed Berlin and presented it as the German capital.

THE MUSEUMSINSEL

In order to become Germany's dignified capital, Berlin also had to become a cultural center. Wilhelm II, who became emperor in 1888, decided to construct a historical area between Lustgarten and Brandenburger Tor, where architecture and the arts would transform Berlin into the "most beautiful city in the world" and proclaim its status as a "world city."[57] National elites worked in close cooperation with the urban bourgeoisie on the enlargement of Berlin's Museumsinsel, or Museum Island.[58] The history of the Museumsinsel had begun in 1825 with the construction of the Altes Museum (Old Museum). In 1841 Emperor Friedrich Wilhelm IV decided to transform the whole island in the Spree into a complex of museums, a place of "arts and science." The Neues Museum opened in 1859, the Nationalgalerie in 1876, the Bodemuseum (formerly the Kaiser-Friedrich-Museum) in 1904, and the Pergamonmuseum in 1930 (construction had begun in 1909). The Museumsinsel reflects the Humboldtian ideal of universal education, and culture and arts from all over the world were exhibited.

From 1881 onward a second museum complex was built to the south of the city, which included the Kunstgewerbemuseum (Museum of Decorative Arts, 1881), the Museum of Ethnology (1886), and the Kunstbibliothek (1867). And even more museums were founded during this period: the Imperial Post Museum (1874), the Natural History Museum (Naturkundemuseum, 1889), the Museum of Pathology (1899), the Museum of Oceanography (1906), and the city museum known as the Märkisches Museum (1907). Thus, in the period between 1870 and 1930, Berlin became one of the foremost museum cities in Europe.[59]

It is striking that the emperor himself attached such importance to the expansion of the Museumsinsel and thus to museums of art, antiquities, and foreign culture. Few museums devoted to science and technology were built—only the Museum of Ethnology, the Natural History Museum, the Museum of Oceanography, and the Museum of Pathology (figure 5.2). Although the Deutsches Museum, a huge and important museum of science and technology, was built in Munich around 1900, no comparably prestigious project can be found in Berlin.[60] In Frankfurt, a museum of natural history (the Senckenbergmuseum) was initiated, set up and financed by the city's bourgeoisie without any help from the state.[61] Yet in Berlin, the urban bourgeoisie showed little interest in the establishment of such institutions. Berlin's scientific museums were either institutions of the Prussian state, such as the Museum of Ethnology, or part of the Friedrich-Wilhelm-University (as were the Museum of Pathology and the Natural History Museum). The Prussian Ministry of Cultural Affairs, along with scientists

FIGURE 5.2
Berlin. Museum of Pathology, Berlin, 1900. Berliner Medizinhistorisches Museum der Charité, Berlin.

such as Rudolf Virchow, the zoologist Wilhelm Peters, and the ethnologist Adolf Bastian, acted as the driving force for the establishment of scientific museums in Berlin, such as the Museum of Pathology and the Natural History Museum. As we will see shortly, the purposes of the museums were manifold: they served to educate the public, to represent the new nation-state, and to legitimize science. History was not their first priority. However, if we look a little more closely, we can see that Virchow's Museum of Pathology and the Natural History Museum both dealt with the past by constructing a story of inevitable progress.

PART VII   SCIENTIFIC MUSEUMS AND THE STORY OF PROGRESS

VIRCHOW'S MUSEUM OF PATHOLOGY

Berlin scientist, physician, and politician Rudolf Virchow opened the Museum of Pathology at the Charité in June 1899. At the time, the museum was the biggest building and the biggest collection of pathological preparations in the world, owning more than twenty thousand preparations when it opened its doors. Contemporaries remarked how extraordinarily large and valuable a collection it was.⁶² The museum covered about 2,000 square meters, of which 600 square meters were open to the public. Plans for such a pathology collection had existed since the early 1830s, because the university's anatomy department already collected specimens.⁶³ However, Virchow was the main protagonist in the process of founding the Museum of Pathology. He set about establishing a pathological collection when he returned to Berlin to accept the professorship for pathology and became the director of the Pathological Institute in 1856. He devoted much energy to the establishment of a Museum of Pathology, which finally became, as he had intended, a part of his Pathological Institute at the Charité. Virchow himself decided almost every detail and was able to carry through his personal ideas on the tasks and objectives of the museum.⁶⁴

The Museum of Pathology had three functions: to collect preparations of diseased parts of the human body, to use them for teaching and research purposes, and to display them to the public. Moreover, the museum must be seen in the context of international competition. Virchow had stressed that the Museum of Pathology was necessary because every hospital school in Great Britain had a collection of pathological preparations.⁶⁵ He felt that Germany had to be competitive with other countries.

When the rector of Berlin's university opened the museum, he stressed the scientific importance of the collection for research and teaching. Virchow himself underlined the need to collect pathological elements, arguing that specialized institutions for collecting were necessary in order to preserve them and to give researchers and students access to them. For almost two centuries doctors had collected pathological preparations at Berlin's university and academy, and thus the museum possessed, as Virchow said, wonderful riches, which had to be protected from neglect or ruin. However, Virchow was not the only one who wanted to have preparations at his disposal. Other doctors tried to obtain organs for their own research. Some even visited autopsies in the Charité to get organs immediately without Virchow's knowledge. Virchow, who spoke of the museum as a "storehouse," complained about such behavior.⁶⁶ He regarded the collection as the only appropriate place for preparations, since it was an "archive of medical knowledge."⁶⁷

Yet for Virchow, the museum's importance was not just scientific; he also recognized the necessity of educating the public by displaying pathological specimens and thereby demonstrating medical progress. Virchow was particularly enthusiastic about using museums as tools for public education. Besides organizing many exhibitions, he was also involved in the founding of the Museum of Ethnology and the Märkisches Museum.[68] In his opening speech at the Museum of Pathology he mentioned his negotiations with the Ministry of Cultural Affairs about opening the museum to the public. He regarded the museum as an important means of preventing disease, in that people could learn about anatomical changes to the body and could learn to recognize diseases and abnormalities. The museum and its displays of specimens and preparations explaining the origin and course of diseases would thus contribute to the scientific understanding and control of the body.[69] The Museum of Pathology was part of Virchow's fight against superstition: he wanted to make clear that the so-called wonders of the human body were explicable by science.[70] His objective was to connect science and life very closely by penetrating daily routines with scientific thinking and behavior.[71]

Five hundred and four hundred visitors, respectively, came to see the collection on the first two Sundays that the museum was open. On the third Sunday, two hundred people visited; thereafter Sundays usually saw between fifty and one hundred visitors. For many visitors, the museum might have meant no more than a chance to look at curiosities and "monsters."[72] Since people have always had many different reasons for visiting museums, it is difficult to ascertain the primary motivation for visitors to the Museum of Pathology. Virchow's successor, Johannes Orth, was initially skeptical toward the idea of opening the museum to the public, and took steps to verify whether people came simply out of curiosity. He finally concluded that visitors were in fact interested in instruction.[73]

However, the dispersal of medical knowledge was not the only reason for Virchow's museum. Virchow wanted visitors not only to learn how to recognize and classify diseases, but also to learn about scientific progress: "The new Museum of Pathology is the result of the collecting activities of the last two hundred years. These activities were extremely important for the progress of medical science, in particular for the development of basic knowledge about the theory of illness." Virchow did not create a continuous narrative in the museum; instead, the display of specimens was presented in a systematic-topological structure, in accordance with the single parts of the body. Nevertheless, he made sure that all visitors to the museum recognized the progress that had occurred in medical science. Simply by showing specimens from over a two-hundred-year span, the museum would, he believed, demonstrate the advance of medical science over those centuries. Virchow had long been aware of the importance of the history of his discipline, and had addressed it in lectures before the museum's opening.[74] He thought of the museum as a place to preserve the memory of doctors and their work, which he regarded as part of cultural history.[75] The objects on display were described as "dankbare Erinnerung" (grateful memory).[76] On every single specimen he noted who the "producer" and collector had been. Thus the specimen was, as Virchow said, a "monument" to the doctor,[77] for each organs, leg, bone, and so on represented the genius of doctors. The museum itself was an "Alben und Gedenkbücher" (memory book).

Without a doubt, Virchow—like his fellow national elites—aimed at inventing tradition. His was the tradition of pathological research, which had started two hundred years earlier, as the collection showed. People could read about medical progress in the newspapers, but Virchow felt that it would be much more effective to *show* progress in the museum. He wanted to demonstrate the current state of the art in medicine in contrast to that of earlier times, when unchecked epidemics such as cholera had killed so many people. The museum displayed, for example, a collection of intestines collected during the cholera epidemic of 1831. Another department showed bones from past wars, starting with the *Befreiungskriege* (wars of liberation) of 1813, followed by the wars against Denmark and Austria, and finally the war against France, which had led to the unification of Germany. Through this collection, the museum illustrated the advances made in treating the wounded over the previous century. This narrative corresponds to the construction of the past in the Siegessäule monument. Virchow's museum and the Siegessäule told related stories of progress: the monument displayed a teleology ending in German unification, while the museum showed how medicine had progressed during the wars that the monument documented. The museum's story of progress helped to legitimize medical research, and to persuade people to believe in modern medicine and to adopt more modern and scientific lifestyles. Virchow's focus on the history of pathology was also part of a strategy to ensure its status as an important discipline, which had to be supported and funded in the future.[78]

NATURAL HISTORY MUSEUM

In 1889 the Museum für Naturkunde was opened, consolidating three existing collections—in mineralogy, paleontology, and zoology—that had been held by departments of Berlin's university since its foundation in 1810.[79] The university's collections had increased enormously as specimens from the colonies flooded into Germany; eventually, lack of space at the university led to demand for a new building. During the Kaiserreich this museum would become Germany's largest and most distinguished natural history collection. The purpose of the museum, as described by the minister of cultural affairs and by scientists such as Virchow, was twofold: it should educate the general public and represent the new nation-state. History was not on the agenda. However, as in the case of the Museum of Pathology, we will see how an idea of "natural" progress influenced the display of the collection.

In 1875 Virchow argued for the establishment of a new "Naturhistorisches Nationalmuseum" in the center of the city. Citing the Austrian museum of natural history as an example, he argued, "If even poor Austria is able to finance and to build such a prestigious building, then the German state must not fall behind. Instead by financing a prestigious museum in the inner city, the German state has to show how the state supports science in Berlin."[80] However, the government did not follow Virchow's advice to the letter. It provided enough money to construct a prestigious building, but not in the center of the city. The Ministry of Cultural Affairs (which financed the building, though it remained under the control of the university) chose a site on the Invalidenstrasse, which was neither at the center of Berlin nor easy to reach by public transportation or even on foot—but the Prussian state owned the area and thus the land did not cost any money. The Ministry of Cultural Affairs never considered placing the new scientific museum at the Museumsinsel, where other museums of arts and culture were located. Wilhelm Peters, director of the zoological collection, complained bitterly because he felt that the location at Berlin's periphery devalued the museum.[81] Since the museum was thought of as part of Germany's national image, the Ministry of Cultural Affairs sought to create a building comparable to famous museums in other European countries. The architect,

August Tiede, traveled to Brussels and Amsterdam, and visited museums of natural history in London, Oxford, Liverpool, and Paris.

The second purpose of the museum was, as mentioned, the education of the general public. Unlike the controversy over making the Museum of Pathology open to the public, the minister of cultural affairs made it very clear that the Natural History Museum had to be a public space. He emphasized that the ministry wanted a museum, not an archive. This was a response to some scientists, such as Peters, curator of the university's zoological collection, who tried to make sure that only a part of the museum would be open to the public, not the whole collection. The minister of cultural affairs decided to invent a new name, "Museum für Naturkunde" (museum for nature study), which is different from "Historisches Museum für Naturkunde" in that it omits the category of history. By emphasizing universal knowledge about nature rather than history, the name is thus a reference to the institution's mission to educate the public, as *Naturkunde* was a subject that was taught in schools.[82]

However, even if "history" was left out of the name of the museum, its collection was structured on evolutionary ideas, and thus by the idea of progress. Since the zoological collection was the largest, we will look more closely at it.[83] Besides being organized according to the theory of evolution, this collection also followed the taxonomic structure of zoology with its hierarchy of class, genus, and species. Kurt Möbius, director of the museum, described the logic of this presentation as descending from the most highly developed animals to the most primitive living beings, shown in inverse order.[84] Entering the museum, the visitor was supposed to follow a prescribed route, beginning in the room where mammals were shown, in which the order of animals started with primates. The next room displayed *vaterländisch* (national) mammals and birds. Then the visitor proceeded through the reptiles, amphibians, and fish, and onward to the insects and spiders. The route ended with the invertebrates, deep-sea animals, and unicellular organisms. As Susanne Köstering has pointed out, the concept by which the museum was structured mirrored the idea of bourgeois society as well as the idea of progress. The taxonomic system, with its classes, families, species, and genera, corresponded with the organization of bourgeois society, whereas the theory of evolution brought in the idea of progress.[85] The museum exhibited a natural development which the visitor experienced in inverse order and which emphasized that the visitor's own place was at the summit of evolutionary development.

PART VIII SHOWING THE FUTURE: BERLIN'S INDUSTRIAL EXHIBITION OF 1896

Berlin's elites reformulated the past in accordance with the needs of the present as well as in accordance with the future they envisioned, since modern concepts of linear time suggested that the future would be the extension of the past and the present. Museums such as Virchow's Museum of Pathology and the Natural History Museum thus not only educated Berlin's residents but also provided spaces in which they could experience the past and future. The paradigmatic modern space for experiencing the future, however, was the universal exhibition.

World exhibitions had many functions—they demonstrated the industrial capabilities of nations and companies, provided a place for the exchange of ideas, and served as festivals of industry—but overall they constituted spaces of experience that provided an orientation toward the future. Since their beginning in 1851, universal exhibitions had been privileged places for the display of innovations. Although, as some contemporaries complained, visitors often preferred the spectacular mass cultural events that accompanied the fairs, organizers viewed world exhibitions as "festivals of progress" that highlighted a vision of the future through brand-new products and the latest technologies. Berlin—a city that was said to have no past, a city which continually reinvented itself—would seem to be the most appropriate place for a world exhibition displaying such a vision of the future. But to this day, no universal exhibition has ever been held in Berlin, though many smaller events have taken place there.

In Germany, the first local exhibitions were held in the 1820s. They served as a means for the state to promote small businesses; thus, in the 1830s and 1840s, when newly emerging associations of small- and medium-sized enterprises started to organize the exhibitions, the number grew enormously. From 1900 until World War I, between 200 and 260 exhibitions were held annually. Exhibitions became, in effect, an important form of mass media.[86]

Five industrial exhibitions were held in Berlin in the nineteenth century. The first took place in 1822, but it was very small, showing only products from the textile industry. The same was the case for the next exhibition in 1827. In 1844, the first large exhibition took place, at which three thousand exhibitors from Berlin's mechanical engineering industry displayed their wares. The next exhibition did not occur until 1879.[87]

In general, exhibitions were regarded as opportunities to present technological and industrial progress. Historian Kenneth W. Luckhurst writes that industrial exhibitions in particular were regarded as an "essentially modern development, closely connected … with the changes, social, economic, political, associated with what is known as the Industrial Revolution."[88] Jürgen Reulecke described the 1844 exhibition as "the most important medium of popularizing the latest economic, technological and scientific knowledge. … Trade and industry exhibitions, in common with few other events, display before the eyes of contemporaries the progress that is so often talked about."[89] This goal is precisely what Berlin's elite hoped to achieve when they proposed that the city host a world exhibition.

The possibility was first discussed after the Paris world exhibition of 1855. The Berlin exhibition of trade and industry in 1879 finally provided the impetus for drawing up plans for a world exhibition, especially since it had attracted many visitors and had been a huge financial success. Immediately afterward its main organizer, the industrialist Fritz Kühnemann, founded an association with the name "1879," in order to prepare and organize a world exhibition.[90] Simultaneously, the Verein Berliner Kaufleute und Industrieller (Association of Berlin Merchants and Industrialists), also founded in 1879, and headed by Ludwig Goldberger, began to promote the idea.[91] And in the words of Julius Lessing, director of Berlin's museum for arts and crafts, Berlin needed a world exhibition because it was a "world city."[92] However, the plans of these Berlin-based elites were strongly opposed by the imperial government and the Centralverband Deutscher Industrieller (Central Confederation of German Industry), who argued that a world exhibition would be much too expensive. Moreover, they asserted that Berlin would not be a good location because it did not lie at an intersection of the world transport system; and, they even added, Berlin had no beautiful landscape around it.[93]

The reasons that the plans for a world exhibition in Berlin failed have been discussed extensively, by contemporaries as well as by more recent scholars.[94] Most scholars suggest that a significant factor was Wilhelm II's strong influence and his decision to prevent Berlin from having a world exhibition. Obviously, he did not like the idea and made no secret of it:

The fame of the Parisians keeps the Berliners awake at night. Berlin is a big city, a world city (perhaps?), and therefore it has also got to have an exhibition! That this argument naturally makes terribly good sense to the proprietors of Berlin's public houses, theaters, dance halls and so forth is not hard to understand. They would be the only ones to profit! Hence all the campaigning. But the *proton pseudos* lies in the assumption that it was tourism alone that brought in all the revenue to Paris. That is not true at all! The hundreds of millionaires who took up residence there for months in order to amuse themselves, attracting ever more acquaintances from the four corners of the earth to join them—they brought in the money. Paris, frankly, is the biggest whorehouse in the world—which Berlin, God grant, will never be—and its appeal stems from that as much as from the exhibition. In Berlin there is nothing to captivate the foreigner beyond the few museums, palaces, and the soldiers.[95]

The emperor's disparaging comments about Berlin and his implicit criticism of the carnival atmosphere of world exhibitions ended with the clear statement that he did not want to have a world exhibition in Berlin "because it is only bad for the city and Germany."[96]

Berlin's tradesmen and industrialists had to accept this decision, even though they had hoped to present to the world the rapid developments in science, technology, and industry that the city had made in recent decades. They sought to present Berlin as a city comparable to Paris, London, or Chicago, but they could not act against the emperor's will. Finally, their grand plans were scaled down to a local exhibition of trade and industry.

The driving forces behind the world exhibition plans—Kühnemann's "1879" association and the Association of Berlin Merchants and Industrialists—turned their attentions to the smaller-scale event. In November 1892 local manufacturers interested in taking part were invited to a meeting to discuss ideas for the proposed exhibition. Kühnemann and Ludwig Max Goldberger, together with members of the local assembly and Bernhard Felisch, a member of the imperial insurance office, constituted the working committee. In all more than one hundred citizens were involved in planning and organizing the exhibition, and seven to eight hundred representatives of Berlin's trade and industry enterprises worked on a voluntary basis in the various subcommittees and commissions. A guarantee fund of 4.5 million marks was set up.[97]

FIGURE 5.3
Berlin. Panorama of Berlin industrial exhibition, 1896, in Treptower Park, contemporary picture postcard. Source: http://de.wikipedia.org/w/index.php?title=Datei:Berlin,_Treptow,_Gewerbeausstellung_1896,_%C3%9Cbersicht.jpg&filetimestamp=20060606231518

The exhibition ran from 1 May to 15 October of 1896 and was held in Treptower Park, which at the time lay on the eastern edge of the city (figure 5.3). Treptow was chosen as the location because it was the biggest space available in Berlin at the time. On the other hand, it was clear from the outset that the site would create problems with traffic—and it did. In the words of one contemporary, "every day endless columns of hackney cabs and horse-drawn carriages traverse the entire city from the western fringes to the Far East." For the masses, a new railroad and the first electric streetcar were built.[98]

Berlin's industrial exhibition was superlative: the site measured some 900,000 square meters, larger than any previous world exhibition held in London, Paris, Vienna, Sydney, or Chicago. Unusually for a local exhibition, all industries in some way connected to Berlin were allowed to participate, regardless of whether the city was their headquarters, whether they had production facilities there, or whether they simply maintained small branch offices. In terms of the number of exhibits, the exhibition was comparable to national exhibitions: About 3,800 exhibitors showed their products.[99] Companies from all key industries found in Berlin participated in the exhibition: textiles, clothing, construction, lumber, porcelain and glass, haberdashery, metallurgy, the graphic and decorative arts, chemicals, food (including alcohol and tobacco), musical and scientific instruments, mechanical engineering, shipbuilding, transport, electrical engineering, leather and rubber, paper, and photography. These were joined by firms involved in sport, health, and welfare facilities, and education.[100]

The trade and industry exhibition also included a colonial exhibition, a reconstruction of old Berlin, and a special amusement park. The exhibition itself was housed in six large buildings and many smaller pavilions.[101] In all there were 7.4 million visitors. National and international newspapers carried reports on the fair; books were written about it, postcards were printed, and an opera was even composed about it.[102] In a commentary on the exhibition, Georg Simmel wrote that it could only be understood within the context of world exhibitions. The abundance and diversity of things shown in the exhibition, he observed, transformed the trade and industry exhibition into an amusement park.[103] In addition, he emphasized its importance for Berlin's self-image as a genuine world city.[104]

One main objective of the exhibition—as the organizers themselves expressed it—was "to make clear how quickly Berlin's industry has developed. ... The Berlin industrial exhibition in 1896 had the purpose of demonstrating to the whole world the great advances made by Berlin's trade and industry, and arts and crafts in recent decades."[105] The exhibition was, without a doubt, a key site of modernity. It showed visitors what it meant to live in a technological world. Not only did it present what great progress Berlin's industry had made in recent decades, but in many respects it opened windows to the future. Theater critic and essayist Alfred Kerr spoke of attending the fair as taking a "pilgrimage to Treptow," for modern technology, he claimed, had become a substitute for religion. The exhibition excited him, and he concluded one article by calling the exhibition "great, great, truly great."[106]

The advances on display were not limited to the spheres of industry and technology. Sections of the exhibition dealt with public health and education, making it clear that Berlin had a modern infrastructure and a "science-based" hygienic lifestyle. Scientists such as Rudolf Virchow and Robert Koch supported these sections. In addition, every day a public lecture was given on science, technology, or modern life. Among the topics were "One day on the moon" and X-ray technology (a lecture at which physicist Conrad Roentgen was present). Aviation pioneer Otto Lilienthal lectured on aircraft, a month before his death in an airplane accident, and Robert Koch gave a talk about the hygienic use of water.[107]

As at the world expositions of this time, people visiting the Berlin exhibition experienced something new: the age of electricity. The electrical industry displayed many of its latest products, including the electric motor and electric cookers.[108] And it presented electricity in impressive ways, supplying the exhibition with light, heating, and power, as well as an electrically lighted fountain. A 3.4-kilometer electric train line was built to connect the main places of interest. Despite the 10-pfennig fee, almost 2.5 million people opted to take the electric train trip.[109] By concentrating on electricity and showing what it was capable of, the exhibition signaled the end of the age of steam.

Although the exhibition poster showed a simple hammer held by a strong male hand erupting from the soil of Brandenburg (an image that some bourgeois contemporaries complained resembled posters of the labor movement), the arts and crafts were scarcely represented at the exhibition.[110] One main reason was that craftsmen could not afford to lease space at the site.[111] Moreover, because the exhibition's aim was to present Berlin as a modern city, its organizers did not mind the lack of craftsmen's participation; the focus on science-based industries made it explicitly clear that people were entering an industrial, modern age, and organizers did not want anything to detract from this atmosphere during the exhibition. The unimportance of handicraft in the coming age was also suggested in a highly symbolic way: in one exhibit, a craftsman was shown with his old, simple hand loom next to a modern machine loom. The superiority of mechanized work was clearly the emphasis.[112]

Berlin's 1896 exhibition was not the first to neglect handicraft. Since Berlin's exhibition of 1844, exhibition organizers had laid down specific requirements that had to be fulfilled by the exhibitors. "Ordinary handicraft which is not produced in big quantities," it was suggested, should be shown only if it was classified as artwork. All exhibits were supposed to meet the needs of a mass society.[113] Thus the exhibition made clear the displacement of handicraft by mass production, and visitors were taught the lesson that the future was one of industrial mass production.

Many items that were displayed were visionary and futuristic; these were probably the most interesting and most spectacular exhibits for ordinary people. Such items included a soda machine and a vending machine for food. Particularly spectacular objects were the baby incubator and the electrical *Stufenbahn*, a type of moving walkway that had already been featured at the Chicago world exhibition. In Berlin it was used to connect the amusement park with the main exhibition area. During the exhibition it was claimed that the *Stufenbahn* would solve all public transport problems in the near future.

As one observer noted, the Berlin exhibition was "halfway between a scientific, technical and industrial demonstration ground and a spectacle and amusement park designed to attract tourism."[114] Critics bemoaned the fact that the exhibition became more of a carnival than an exhibition of trade and industry.[115] "Industry," wrote one reporter, "is only the necessary evil which does not attract attention anymore."[116] However, even in this respect the exhibition reflected an important part of modern life, as was illustrated in Simmel's comment that exhibits were not only shown in their functional guise but were "staged" and presented as "temptations," thus underlining the "shop-window quality" of the exhibits. The exhibition, according to Simmel, was a new "synthesis between the outer attractiveness and the actual functionality of the objects."[117] Their diversity and abundance, in his view, excited the nerves and satisfied man's need for excitement.[118] But at the same time, Simmel cautioned, the abundance of exhibits led to a "paralysis of one's faculty of perception" and a state of "genuine hypnosis" in which everything was reduced to the level of entertainment.[119] That assessment was especially true of the many festivals, concerts, and fireworks displays accompanying the exhibition, including the First German Colonial Exhibition and the "Old Berlin" and "Cairo" amusement parks. "Old Berlin" was a reconstruction of Berlin as it had existed between 1650 and 1665.[120] "Cairo" similarly imitated pyramids, Cairo's old town, a labyrinth, a bazaar, and a harem.[121]

The Berlin exhibition was an ephemeral event, intended from the beginning to be temporary.[122] By contrast with many other exhibitions, very few traces remained once it closed in October 1896—in fact, Berlin's municipal authority had agreed to the use of Treptower Park only on condition that no traces of the exhibition should be left. The buildings were taken down, the artificial lake filled up with soil. What was left was a much-trampled park whose regeneration took years, along with a huge telescope, a newly built bridge (Oberbaumbrücke), and the electric streetcar.[123] Berlin's inhabitants now had two new public transport routes for excursions to Treptower Park. The new lines encouraged the overall electrification of Berlin's transport system, which was completed in 1902.[124]

In the end the exhibition did not have the hoped-for financial success. Two million German marks had to be taken from the guarantee fund.[125] The project's critics, who had argued from the outset that a world exhibition would be a financial disaster, felt their opposition had been justified by visitors' enthusiasm for spectacle and amusement. However, in 1904 some influential inhabitants of Berlin, including Emil Rathenau, Adolf Slaby, and Max Goldberger, again raised the question of holding a world exhibition in Berlin. But, once again, the emperor denied their request.[126]

PART IX   CONCLUSION

Berlin's change from a backward town in remote Brandenburg into a modern, science-based city was rapid and took place on a tremendous scale. It is striking that the change was not driven by the emperor or the bourgeois elites. Rather, a network of scientists, engineers, entrepreneurs, and politicians were the driving force for the founding of scientific institutions, the support of technological progress, and the remaking of Berlin as a modern, hygienic city fostering a scientific lifestyle. In particular, Rudolf Virchow acted as a kind of node by virtue of belonging to different networks. Some of the protagonists of the culture of change, such as Virchow, were both scientists and politicians; others, such as Werner von Siemens, were both engineers and entrepreneurs. These actors established a culture of change, which emphasized the necessity of creating a present according to scientific-technological rationality, oriented toward the future, with the past employed as a tool to achieve this end. They were concerned with constructing the present by changing the urban environment, by founding new institutions and establishing new urban technologies, and by spreading scientific knowledge and a rational, scientific lifestyle. Believing in science and technology as the only appropriate solution for social and hygienic problems, they held the main objective of establishing an industrial, scientific way of life. They also reformulated the past by inventing a tradition of scientific progress.

And they opened windows into the future by establishing museums and holding Berlin's industrial exhibition at the end of the nineteenth century.

Although Emperor Wilhelm II was famous for his interest in science and technology, he did not do much to transform Berlin into a science-based city. On the one hand he regarded institutions such as the Technical University with favor, he listened to lectures of famous scientists, and he supported the founding of the Kaiser-Wilhelm-Institutes. On the other hand he refused to host a world exhibition in Berlin, and hindered the establishment of innovations such as the electric streetcar and the subway, if they came too close to important historical buildings or prestigious streets such as Unter den Linden. He was much more interested in transforming Berlin into a Prussian city or an arts city than a science-based city. Whenever decisions about Berlin's urban development were contested, Wilhelm II decided in favor of aesthetics and politics instead of science and technology.[127]

How did Berlin's inhabitants react to these efforts to transform Berlin into a modern metropolis? Unfortunately, as is often the case, we do not have the same wealth of sources from this quarter as we have from officials and scientists. However, contemporaries made numerous comments suggesting that Berlin's inhabitants behaved in a very traditional way. As mentioned above, Anthony Sutcliffe has described Berlin as a metropolis whose inhabitants were not ready for the process of modernization. Doctors reported that people still preferred the chamber pot to the water closet. And the socialist Eduard Bernstein described late-nineteenth-century Berlin residents as still holding rural political convictions and judgments.[128] Thus, the process of transforming Berlin into a modern city was simultaneously a huge education project, undertaken in order to transform rural people into modern city dwellers.

Compared with other metropolises such as London, Paris, and Chicago, Berlin was a latecomer. Yet, given the city's backwardness in the early nineteenth century, its subsequent degree of change was tremendous. Berlin wanted to be a world city, and emulation and competition were important motivations among its elites. The founding of museums and the never-realized plans for a world exhibition must be seen in the context of emulation. As we've seen, James Hobrecht traveled to other cities to inform himself about urban planning and sewage systems. Haussmann was an ever-present model in Berlin, while Berlin's sewage system, in turn, became a model for other cities. Virchow looked to British hospitals when planning his Museum of Pathology, and he wanted Berlin's Natural History Museum to surpass Austria's efforts. As these examples illustrate, the emergence of an international culture of change in Berlin can only be understood within the context of a network of competing cities.

NOTES

1. Karl Scheffler, *Berlin, ein Stadtschicksal* (Berlin: Reiss, 1910), 266f.

2. Ibid.

3. Constantin Goschler, "Wissenschaft und Öffentlichkeit in Berlin (1870–1930): Einleitung," in Constantin Goschler, ed., *Wissenschaft und Öffentlichkeit in Berlin, 1870–1930* (Stuttgart: Steiner, 2000), 15; see also 7–29.

4. Laurenz Demps, "Von der preussischen Residenzstadt zur hauptstädtischen Metropole," in Werner Süss and Ralf Rytlewski, eds., *Berlin. Die Hauptstadt. Vergangenheit und Zukunft einer europäischen Metropole* (Berlin: Nicolai, 1999), 17–51.

5. Eric Hobsbawm and Terrence Ranger, *The Invention of Tradition* (Cambridge: Cambridge University Press, 1983).

6. David Gugerli, "Modernität-Elektrotechnik-Forschritt: Zur soziotechnischen Semantik moderner Erwartungshorizonte in der Schweiz," in Klaus Plitzner, ed., *Elektrizität in der Geistesgeschichte* (Stuttgart: GNT-Verlag, 1998), 67.

7. P. Clausewitz, "Berlin um Jahr 1806," quoted in Jochen Boberg, Tilman Fichter, and Eckhart Gillen, eds., *Experimentierfeld der Moderne. Industriekultur in Berlin im 19. Jahrhundert* (Munich: Beck, 1984).

8. Cited in Benedikt Goebel, *Der Umbau Alt-Berlins zum modernen Stadtzentrum. Planungs-, Bau- und Besitzgeschichte des historischen Berliner Stadtkerns im 19. und 20. Jahrhundert* (Berlin: Braun, 2003), 40.

9. Gerhard Brunn, "Einleitung: Metropolis Berlin. Europäische Hauptstädte im Vergleich," in Gerhard Brunn and Jürgen Reulecke, eds., *Metropolis Berlin. Berlin als deutsche Hauptstadt im Vergleich europäischer Hauptstädte 1871–1939* (Bonn and Berlin: Bouvier, 1992), 34.

10. Demps, "Von der preussischen Residenzstadt zur hauptstädtischen Metropole," 44.

11. Cited in Werner Siebel, "Fabrikarbeit und Rationalisierung," in Boberg, Fichter, and Gillen, *Exerzierfeld der Moderne*, 311; see also 310–312.

12. Cited in Heinz Reif, "Hauptstadtentwicklung und Elitenbildung: 'Tout Berlin' 1871 bis 1918," in Michael Grüttner, Rüdiger Hachtmann, and Heinz-Gerhard Haupt, eds., *Geschichte und Emanzipation* (Frankfurt: Campus, 1999), 685; see also 679–699.

13. August Bebel, *Aus meinem Leben*, vol. 2 (Berlin: Dietz, 1953), 125.

14. Goschler, "Wissenschaft und Öffentlichkeit," 15. Compare Marc Schalenberg and Rüdiger vom Bruch, "London, Paris, Berlin. Drei wissenschaftliche Zentren des frühen 19. Jahrhunderts im Vergleich," in Richard van Dülmen and Sina Rauschenbach, eds., *Macht des Wissens* (Cologne: Böhlau, 2004), 681–699.

15. Goschler, "Wissenschaft und Öffentlichkeit," 20.

16. Hubert Laitko, *Wissenschaft in Berlin: Von den Anfängen bis zum Neubeginn nach 1945* (East Berlin: Akademie Verlag, 1987), p. 321.

17. Brunn, "Einleitung," 13.

18. Günter Richter, "Zwischen Revolution und Reichsgründung (1848–1870)," in Wolfgang Ribbe, ed., *Geschichte Berlins*, vol. 2, *Von der Märzrevolution zur Gegenwart* (Munich: Beck, 1987), 657; see also 605–687.

19. David Cahan, *An Institute for an Empire: The Physikalisch-Technische Reichsanstalt 1871–1918* (Cambridge: Cambridge University Press, 1989); and Bernhard vom Brocke, "Die Kaiser-Wilhelm-Gesellschaft im Kaiserreich," in Rudolf Vierhaus and Bernhard vom Brocke, eds., *Forschung im Spannungsfeld von Politik und Gesellschaft* (Stuttgart: Deutsche Verlagsanstalt, 1990), 17–162.

20. Sigfrid von Weiher, *Berlins Weg zur Elektropolis. Technik- und Industriegeschichte an der Spree* (Berlin: Stapp, 1974).

21. Cahan, *An Institute for an Empire*, 24.

22. Laitko, *Wissenschaft in Berlin*, 248.

23. Ibid., 222.

24. Mitchel Ash, "Wissenschaftspopularisierung und Bürgerliche Kultur im 19. Jahrhundert. Essay-Rezension," *Geschichte und Gesellschaft* 28 (2002): 322–334.

25. Laitko, *Wissenschaft in Berlin*, 300.

26. Jochen Boberg, Tilman Fichter, and Eckhart Gillen, "Einleitung, Jungalte Stadt," in Boberg, Fichter, and Gillen, *Experimentierfeld der Moderne*, 7, 6–9.

27. Gottfried Korff, "Berliner Nächter: Zum Selbstbild urbane Eigenschaften und Leidenschaften," in Brunn and Reulecke, *Metropolis Berlin*, 71–103.

28. Berlin served as a model for Stockholm and Helsinki, for example; see Marjetta Hietala, "Berlin und andere deutsche Städte als Vorbild für die Stadtverwaltung

von Helsinki und Stockholm um die Jahrhundertwende," in Brunn and Reulecke, *Metropolis Berlin,* 201–223.

29. Klaus Strohmeyer, *James Hobrecht (1825–1902) und die Modernisierung der Stadt* (Potsdam: Verlag für Berlin-Brandenburg, 2000). See also Richard Evans, *Tod in Hamburg. Stadt, Gesellschaft und Politik in den Cholara-Jahren 1830–1910* (Reinbek bei Hamburg: Rowohlt, 1996).

30. Cited in Dieter Schott, "Wege zur vernetzten Stadt: Technische Infrastrukturen in der Stadt aus historischer Perspektive," *Information zur Raumentwicklung* 5 (2006): 249; see also 249–257.

31. Ibid., 250.

32. Karl Schwarz, ed., *Von der Residenzstadt zur Industriemetropole*, vol. 1, *Aufsätze* (Berlin: Kompass, 1981), 75.

33. Cited in Constantin Goschler, *Rudolf Virchow: Mediziner, Anthropologe, Politiker* (Cologne Weimar, and Vienna: Böhlau, 2002), 250.

34. Ibid.

35. Ibid., 249.

36. Ibid.

37. Horst Siewert, "Die Stadtbahnprojekte," in *Experimentierfeld*, 102, see also 102–105.

38. Goschler, *Rudolf Virchow*, 251.

39. Strohmeyer, *James Hobrecht*, 18.

40. In particular, Karl Scheffler and Werner Hegemann were prominent critics of the Hobrecht Plan: see Scheffler, *Berlin*; Werner Hegemann, *Das steinerne Berlin* (Berlin: Kiepenheuer, 1930).

41. Johann Friedrich Geist and Klaus Kürvers, *Das Berliner Mietshaus*, vol. 2 (Munich: Prestel, 1984).

42. Strohmeyer, *James Hobrecht*, 60.

43. Ibid., 53.

44. Ribbe, *Geschichte Berlins*, 229.

45. Goschler, *Rudolf Virchow*, 251.

46. Rudolf Virchow, *Reinigung und Entwässerung Berlin* (Berlin, 1873).

47. Cf. Goschler, *Rudolf Virchow*.

48. Ibid., 251f.

49. Ibid., 252ff.

50. Scheffler, *Berlin*.

51. Cécile Chombard-Gaudin, "Frankreich blickt auf Berlin 1900–1930," in Brunn and Reulecke, *Metropolis Berlin*, 367–407.

52. Cited in Ruth Glatzer, ed., *Das Wilhelminische Berlin: Panorama einer Epoche, 1890–1918* (Berlin: Siedler, 1997), 39.

53. Benedikt Goebel scrutinized this process very thoroughly; see Goebel, *Umbau*.

54. Hans J. Reichhardt, "Stadterweiterung," in Boberg, Fichter, and Gillen, *Exerzierfeld der Moderne*, 96; see also 90–97.

55. Glatzer, *Das Wilhelminische Berlin*, 51.

56. Matthias Braun, *Die Siegessäule* (Berlin: Berlin Edition, 2000); and Bernd Sösemann, "Exerzierfeld und Labor deutscher Geschichte. Berlin im Wandel der deutschen und europäischen Politik zwischen 1848 und1933," in Süss and Rytlewski, *Berlin*, 100–122, 107ff.

57. Glatzer, *Das Wilhelminische Berlin*, 42

58. Thomas W. Gaehtgens, *Die Berliner Museumsinsel im Deutschen Kaiserreich* (Munich: Deutscher Kunstverlag, 1992).

59. Ribbe, *Geschichte Berlins*, 788.

60. Oskar von Miller was closely involved in the foundation of the Deutsches Museum. When he visited the Conservatoire des Arts et Métiers in Paris, South Kensington Museum in London, and the Urania in Berlin, he started thinking about creating a similar German museum. He managed to get support for his idea not only from the Bavarian government but also from Emperor Wilhelm II. The Deutsches Museum was built to give an encyclopedic overview of all areas of science and technology and their history. The historical exhibits, called *Meisterwerke* (masterpieces), were intended to emphasize the scientific and technological achievements of individuals as well as to give a general view of progress in these areas.

61. In 1907 the Senckenbergmuseum was reopened in a new and impressive building, which was also financed by Frankfurt's bourgeoisie. See Susanne Köstering, *Natur zum Anschauen: Das Naturkundemuseum des deutschen Kaiserreichs 1871–1914* (Cologne, Weimar, and Vienna: Böhlau, 2003), 34.

62. Peter Krietsch and Manfred Diestel, *Pathologisch-Anatomisches Cabinet. Vom Virchow-Museum zum Berliner Medizinhistorischen Museum in der Charité* (Berlin and Vienna: Blackwell, 1996), 1.

63. Ibid., 17, 19.

64. Angela Matyssek, "Die Wissenschaft als Religion, das Präparat als Reliquie," in Ante te Heesen and E. C. Sarpy, eds., *Sammeln als Wissen. Das Sammeln und seine wissenschaftsgeschichtliche Bedeutung* (Göttingen: Wallstein, 2002), 146; see also 142–163.

65. Angela Matyssek, *Rudolf Virchow. Das Pathologische Museum. Geschichte einer wissenschaftlichen Sammlung um 1900* (Darmstadt: Steinkopf, 2002), 38.

66. Rudolf Virchow, *Die Eröffnung des Pathologischen Museums der Königlichen Friedrich-Wilhelms-Universität zu Berlin am 27. Juni 1899* (Berlin 1899), 16f.

67. Oskar Israel, "Das Pathologische Museum der Königlichen Friedrich-Wilhelms-Universität zu Berlin," *Berliner Klinische Wochenschrift* 41 (1901): 1051; see also 1047–1052.

68. Goschler, *Rudolf Virchow*, 263; see also Matyssek, *Rudolf Virchow*, 33.

69. Matyssek, *Rudolf Virchow*, 25.

70. Ibid.

71. Ibid., 35.

72. Ibid., 44.

73. Johannes Orth, "Das Pathologische Institut in Berlin," in Johannes Orth, ed., *Arbeiten aus dem Pathologischen Institut zu Berlin. Zur Feier der Vollendung der Institutsneubauten* (Berlin, 1906), 20; see also 1–176.

74. Matyssek, *Rudolf Virchow*, 50.

75. Matyssek, "Die Wissenschaft als Religion," 39.

76. Matyssek, *Rudolf Virchow*, 26.

77. Cited in Matyssek, "Die Wissenschaft als Religion," 50.

78. Matyssek, *Rudolf Virchow*, 42f.

79. More details can be found in Carsten Kretschmann, *Räume öffnen sich: Naturhistorische Museen im Deutschland des 19. Jahrhunderts* (Berlin: Akademie Verlag, 2006), 30ff.

80. Cited in Köstering, *Natur zum Anschauen*, 50.

81. Kretschmann, *Räume öffnen sich*, 37.

82. Historische Bild- und Schriftgutsammlung, file Verwaltungsakten, SII, Archive of Museum für Naturkunde der HU zu Berlin.

83. Cf. *Führer durch die zoologische Schausammlung des Museums für Naturkunde in Berlin* (Berlin, 1899).

84. Kommentar Möbius, Verwaltungsakte, Archive of Museum für Naturkunde der HU zu Berlin.

85. Köstering, *Natur zum Anschauen*, 277f.

86. Thomas Grossbölting, "Die Ordnung der Wirtschaft. Kulturelle Repräsentationen in den deutschen Industrie- und Gewerbeausstellungen des 19. Jahrhunderts," in Hartmut Berghoff and Jakob Vogel, eds., *Wirtschaftsgeschichte als Kulturgeschichte. Dimensionen eines Perspektivwechsels* (Frankfurt: Campus, 2004), 382; see also 377–404.

87. Cf. Paul Thiel, "Berlin präsentiert sich der Welt. Die Treptower Gewerbeausstellung 1896," in Boberg, Fichter, and Gillen, *Exerzierfeld der Moderne*, 16–27.

88. Kenneth W. Luckhurst, *The Story of Exhibitions* (London: Studio Publications, 1951), 14.

89. Jürgen Reulecke, *Sozialer Frieden durch soziale Reform. Der Centralverein für das Wohl der arbeitenden Klassen in der Frühindustrialisierung* (Wuppertal: Hammer, 1983), 36.

90. Elke Reuter, "Die Grosse Berliner Gewerbe-Ausstellung 1896 im Treptower Park," *Berlinische Monatsschrift Luisenstadt* 1 (1992): 5, see also 4–10.

91. Ibid., 5.

92. Alexander Geppert, "Deutsche Grossausstellungsprojekte und ihr Scheitern, 1880–1930," *Wolkenkuckucksheim* 5 (2000): 1, 4f.

93. Reuter, "Grosse Berliner Gewerbe-Ausstellung," 5.

94. Cf. Geppert, "Deutsche Grossausstellungsprojekte," 15.

95. Cited in Glatzer, *Das Wilhelminische Berlin*, 90.

96. Ibid.

97. Reuter, "Grosse Berliner Gewerbe-Ausstellung," 7.

98. Max Osborn, *Berlins Aufstieg zur Weltstadt*, cited in Glatzer, *Das Wilhelminische Berlin*, 90.

99. Grossbölting, "Die Ordnung der Wirtschaft," 380f.

100. Erhard Crome, "Berliner Gewerbeausstellung 1896. Betrachtung eines Jahrhundertstücks," in *Die verhinderte Gewerbeausstellung. Beiträge zur Berliner Gewerbeausstellung 1896* (Berlin: Berliner Debatte Wissenschaftsverlag, 1996), 24; see also 11–28.

101. Reuter, "Grosse Berliner Gewerbe-Ausstellung," 9.

102. Grossbölting, "Die Ordnung der Wirtschaft," 381.

103. Georg Simmel, "Berliner Gewerbeausstellung," *Die Woche* (15 July 1896): 59.

104. Ibid.

105. Wilhelm C. Bacharach, "Das Propaganda-Bureau," in Arbeits-Ausschuss, ed., *Berlin und seine Arbeit. Amtlicher Bericht der Berliner Gewerbe-Ausstellung 1896 zugleich eine Darstellung des gegenwärtigen Standes unserer gewerblichen Entwicklung* (Berlin, 1898), 88.

106. Alfred Kerr, *Mein Berlin. Schauplätze einer Metropole* (Berlin: Aufbau-Verlag, 1999), 126.

107. Crome, "Berliner Gewerbeausstellung 1896," 25.

108. Uwe Müller and Frank Zschaler, "Weltstadt von Industrie und Gewerbe— Die Berliner Wirtschaft und die Gewerbeausstellung von 1896," in *Die verhinderte Gewerbeausstellung*, 45; see also 29–55.

109. Ibid., 41f.

110. Cf. Hellmut Rademacher, "Auf dem Weg zum künstlerischen Plakat. Ludwig Sütterlins Entwurf zur Berliner Gewerbeausstellung," in *Die verhinderte Gewerbeausstellung*, 97; see also 97–103.

111. Hans-Heinrich Müller, "Eine Parade der Produktion. Die Berliner Gewerbeausstellung von 1896," *Berlinische Monatsschrift* 5, no. 4 (1996): 32; see also 31–35.

112. Siebel, "Fabrikarbeit und Rationalisierung," 312.

113. Grossbölting, "Die Ordnung der Wirtschaft," 386.

114. Quotation from the period, in Müller, "Eine Parade der Produktion," 35.

115. Reuter, "Grosse Berliner Gewerbe-Ausstellung," 8.

116. *Die Zukunft*, 4 July 1896.

117. Simmel, "Berliner Gewerbeausstellung," 60.

118. Ibid., 59.

119. Ibid.

120. Reuter, "Grosse Berliner Gewerbe-Ausstellung," 8

121. Ibid.

122. Crome, "Berliner Gewerbeausstellung 1896," 23.

123. Müller, "Eine Parade der Produktion," 35.

124. Reuter, "Grosse Berliner Gewerbe-Ausstellung," 9.

125. Ibid., 10.

126. Wolfgang König, *Wilhelm II. und die Moderne* (Munich, Vienna, and Zurich: Ferdinand Schöningh, 2007), 150f.

127. Ibid., 265.

128. Cited in Siebel, "Fabrikarbeit und Rationalisierung," 311.

6 PROMOTING SCIENTIFIC AND TECHNOLOGICAL CHANGE IN TOKYO, 1870–1930: MUSEUMS, INDUSTRIAL EXHIBITIONS, AND THE CITY

MORRIS LOW

Everything looked as though it were being destroyed, and at the same time everything looked as though it were under construction. To Sanshirō, all this movement was horrible. His shock was identical in quality and degree to that of the most ordinary country boy who stands in the midst of the capital for the first time.

—Natsume Sōseki, *Sanshirō* (1908)

INTRODUCTION

In the novel *Sanshirō* (1908), which traces the life of Ogawa Sanshirō, a twenty-three-year-old student at Tokyo Imperial University, the well-known Japanese novelist Natsume Sōseki[1] (1867–1916) provides a window to understanding what life was like in Japan's capital during the Meiji period (1868–1912). Newly arrived—by ferry and train—from Fukuoka prefecture on the island of Kyūshū, Sanshirō is startled by the size of Tokyo and the continual stream of streetcars.[2]

Indeed, the railway helped link everyone in Japan more closely to Tokyo, where they would find a chaotic montage of the old and new. An old temple could be found sitting next to a Western-style house, painted blue. In the novel, against the backdrop of a fast-modernizing city, Sanshirō encounters violence when a young woman commits suicide by jumping in front of a train and he sees the remains of her body left on the track. To the young man, already shocked by the city's constant change, the suicide suggests the brutality of modern technology.

Sōseki wrote from personal experience. He became familiar with some of the modern institutions when he was sent overseas on a Japanese government scholarship to learn from the West. He visited England from 28 October 1900 through 5 December 1902, during which time he studied

English literature at University College London and visited various museums, including the British Museum, the South Kensington Museum (the forerunner of the Victoria and Albert Museum and the Science Museum), the Royal Academy of Art, and the Wallace Collection.[3] He thus came to understand the achievements of Western culture and at the same time all the ills that a modern metropolis and city life involved. Sōseki came to view Japan's rushed modernization as unnatural. The Japanese had only focused on importing the surface culture of the West, and he believed that the nation of Japan, like himself, was bordering on having a nervous breakdown.[4] His work helps us to understand how problematic the rapid cultural change during the Meiji era was for the Japanese people. This chapter examines some of the institutions that were introduced into Japan and key figures behind the changes that took place.

More than a century later, Tokyo remains a vibrant, attractive, modern city. Not only is it Japan's capital, but it is "the only city in Japan with real clout."[5] It remains the center of scientific, cultural, and economic activity, with some 59 percent of major Japanese corporate headquarters there. Tokyo's ability to reinvent itself and encourage change has been integral to its success. How has Tokyo been so resilient?

The first part of this chapter examines the reinvention of Tokyo as imperial capital during the Meiji era. The changes that Tokyo experienced at this time—in which science and technology were integral—were synonymous with the greater transformations that the nation underwent. If Japan was to modernize, it was crucial that Tokyo take the lead. This urban development was seen by the poet Hagiwara Sakutarō (1886–1942) and others as a metaphor for modernity and the West.[6]

In contrast to China, Japan was more willing to borrow and learn from other cultures in the late nineteenth century. How were the Japanese able to promote a culture of change when the Chinese were not able to do so? Part II examines how looking to the past was necessary in order for Japan to move forward. Museums were important in this process. Participation in international exhibitions and the holding of domestic industrial exhibitions also helped the Japanese to envision a future led by science and technology.

To revise unequal treaties with Western powers that were signed in the aftermath of Commodore Matthew Perry's visits to Japan in 1853 and 1854, Japan had to show the rest of the world that it was making progress.

Part III focuses on how Japan put itself on display abroad, and on the educational value of showcasing its achievements and foreign know-how at home.

But natural disasters and a lack of political commitment sometimes interfered with modernization efforts. The fourth part of the chapter examines how Gotō Shinpei (1857–1929), mayor of Tokyo and then home minister, sought to rebuild Tokyo according to "scientific" principles after the Great Earthquake of 1923. He was ultimately frustrated and his dreams for Tokyo were not fully realized. Instead, he looked to the growing Japanese empire for opportunities to realize his vision.

PART I   MAKING A MODERN CITY

In the Meiji era (1868–1912), Japan embarked on an ambitious program of rapid modernization, a feature of which was the large-scale introduction of Western science and technology. There were great efforts to build a modern technological urban environment that included the promotion of scientific and technical education, the establishment of research organizations and museums, and the founding of industrial enterprises.

HOW BIG WAS TOKYO?

What sort of place was Tokyo in the nineteenth century? Although Edo (present-day Tokyo) had humble beginnings, having developed from a number of fishing villages around a castle, by the eighteenth century it was one of the largest cities in the world. Even Kyoto and Osaka had populations in the hundreds of thousands, making them larger than Vienna, Moscow, and Berlin at beginning of the nineteenth century. The population of Edo, estimated at some 600,000 during much of the Tokugawa period (1603–1868), rivaled that of Beijing. By the late eighteenth century, Japan had 3 percent of the world's population, but more than 8 percent of the world's population living in cities of more than 10,000 people. Ten percent of the Japanese population in 1800 can be said to have been urban.[7]

Shogunates, a type of military dictatorship, had existed in Japan since the establishment of a feudal government in the twelfth century. The emperor and his court were relegated to an obscure life in Kyoto, unable to wield political power. Tokugawa Ieyasu (1542–1616) established the last shogunate in the early seventeenth century and began what is known as the Tokugawa period (also called the Edo period because of the location of the shogun's capital at Edo, present-day Tokyo).[8]

Edo castle and the surrounding city became the national center of power and influence, supplanting the ancient capital of Kyoto where the imperial family resided. Although there had been plans in 1602 to develop the city along the lines of a four-directional grid similar to that of Kyoto, it was decided instead that the city should grow by extending canals outward from the castle at its center in a spiral fashion. This development was superimposed on the earlier plan of five major roads which radiated out from the center. In this way, major thoroughfares were maintained and civil engineers were able to continue expanding the city indefinitely by enlarging the canal system.[9]

Edo was a large but clean and healthy city. The *sankin kōtai* system of alternate residence—whereby feudal lords were forced to come to Edo, take up residence, and leave their families there as virtual hostages when they returned to the provinces—caused the city to grow and triggered growth in production throughout Japan. But despite this growth, Susan B. Hanley has argued that metropolitan sanitation was better in Japan from the mid-seventeenth century to mid-nineteenth century than in the West, in terms of water supply and waste disposal.[10] The population and mortality rates testify to the fact that a healthier environment was created for urban populations. Customs concerning hygiene, food, and drink, combined with the lack of domestic animals, led Japanese city life to be more sanitary than urban life in the West.[11]

THE MILITARY IMPERATIVE BEHIND MODERNIZATION

The expedition of Matthew Perry in 1853–1854 effectively opened up Japan to Western trade. Commodore Perry was sent to Japan to ensure protection of American seamen, to gain access to Japanese ports for provisions and coal, and to seek rights of trade. A treaty was signed on 31 March 1854, at Kanagawa, in which the Japanese agreed to immediately open up the port of Shimoda and to open the port of Hakodate in a year's time. Similar agreements were soon signed with Russia and Britain.[12]

Townsend Harris, the first U.S. consul in Japan, subsequently negotiated the first major commercial treaty between Japan and a Western power in the summer of 1858.[13] About two weeks after the signing of the treaty,

Lord Elgin arrived in Edo and by 26 August had signed a treaty based on Harris's. This had been preceded a few days earlier by agreements with Holland and Russia, and was followed in October by an agreement with France. With the signing of these "unequal treaties," Japan entered into full commercial relations with five Western powers.[14] Between 1854 and 1873, treaties were signed with a total of sixteen Western nations: Austria-Hungary, Belgium, Denmark, France, Germany, Great Britain, Hawaii, Italy, The Netherlands, Peru, Portugal, Russia, Spain, Sweden-Norway, Switzerland, and the United States.[15]

Despite substantial economic growth and urbanization during the Tokugawa period, the beginning of Japan's modernization is generally seen as beginning with the Meiji Restoration of 1867–1868, in which the emperor was restored to power. In the years leading to the Meiji Restoration, antiforeign feeling had been aggravated by the trade agreements, and the flood of foreign manufactured goods—made possible by the limits on tariffs imposed by the unequal treaty system—had led to the disintegration of the economy. Because of the economic distress of the peasants and lower samurai, antagonism against the feudal government and foreigners grew, leading to agrarian revolts and frequent assassinations of top feudal officials. Ultimately the Tokugawa Shogunate was overthrown, and the emperor was restored to direct rule of the country.[16]

With the Meiji Restoration, the feudal government (*bakufu*) was overthrown by lower samurai and *rōnin* (masterless samurai) from the western clans of Satsuma, Chōshū, Tosa, and Hizen, supported by merchants—especially those from Osaka, such as Mitsui—in whom 70 percent of the nation's wealth had come to be concentrated. The Restoration was effectively a shift of governmental power from the upper to the lower samurai, facilitated by the merchants.[17] Key figures from these groups played an important role in the modernization of Japan with Tokyo as its capital.

MEIJI MODERNIZATION

The Meiji government devoted its energy to centralization and modernization of the army and police force. A modern army and navy required strategic industries such as heavy industries, engineering, mining, and shipbuilding. Military industries had already been introduced by the Satsuma, Hizen, and Chōshū clans before the Restoration. The Meiji government took the lead in mining and heavy industrial production. Engineering, technical, and naval schools were established with the assistance of foreign instructors, and students were also sent abroad to master technology in key industries. Transportation and communication were developed as well.[18]

In April 1868, a Charter Oath was issued which provided a framework for the policy of the new government. In 1871, the domains, which had been ruled by clans, were abolished and replaced by prefectures, governed by the emperor's appointees. Edo became the imperial capital.[19]

The beginning of the Meiji period was characterized by the emergence of new organizations that were modeled on those of the major Western powers. These took the form of new national systems, including central and local administration; primary schools; police; postal communications and telegraph offices; law courts; army and navy; and commercial organization such as banks, factories, newspapers, railways, chambers of commerce, and joint stock companies.[20] Many of these new organizations were based in Tokyo.

It was important to rally the Japanese people behind the program of modernization, and crucial to show the rest of the world that Japan was a civilized nation that had achieved much and was destined for bigger and better things. Much of the borrowing of models took place in the 1870s, the first decade of the Meiji period, and was promoted by slogans such as *fukoku kyōhei* (enrich the country, strengthen the army) and *bunmei kaika* (civilization and enlightenment). The program of modernization was motivated by the major goals of building military capability equal to that of the Western powers and revising the unequal treaties that had been signed by the Tokugawa Shogunate and reaffirmed by the new Meiji government.

A combination of government officials who had been samurai and a rising class of modern business entrepreneurs propelled the modern growth of the Meiji economy.[21] The Meiji government made many entrepreneurial decisions, as did the *zaibatsu* (large, family-controlled financial combines that came to dominate Japan's economy). The Department of Home Affairs, under Ōkubo Toshimichi (1830–1878), is regarded as being the center of the government's industrial action in the market. The *zaibatsu*, especially Mitsui, Mitsubishi, Sumitomo, and Yasuda, provided the capital for the activities of Meiji entrepreneurs.[22] The success of the entrepreneurs depended to a large extent on the banks as sources of capital. A key figure,

Shibusawa Eiichi, was president of the Dai Ichi Ginkō (First National Bank).[23]

Itō Hirobumi (1841–1909) and Yamao Yōzō also played leading roles in the technological transformation of Tokyo by founding the Ministry of Public Works. They had become convinced of the need for modernization after visiting Britain secretly on a study tour in 1863 along with three other young men from the Chōshū domain. It is no coincidence that it was lower-ranking samurai from the Chōshū domain who helped overthrow the Tokugawa regime in 1868.[24] In December 1870, Itō and Yamao established the Ministry of Public Works with support from colleagues such as Ōkubo Toshimichi. The ministry, which facilitated the introduction of Western technology, consisted of ten departments: engineering, promotion of industry, mining, railways, civil engineering, lighthouse construction, shipbuilding, telegraphy, iron-making, and manufacturing. Its achievements included the construction of a railway from Tokyo to Yokohama in 1872, and the creation of a nationwide telegraph network. It was, in many ways, a learning organization, with Japanese engineers undergoing training at various work sites, helped by foreigners employed by the government, who were known as *oyatoi gaikokujin* or "honorable, foreign hirelings."[25]

In any year of the Meiji period there were around eight thousand foreigners employed throughout Japan. Half of these were Chinese day laborers; only about three thousand were professionals in government service, who tended to work in Tokyo. The peak period was 1874 and 1875, with eight hundred new foreign employees hired in each of these years. The average length of service of professional advisors was around five years, and nine years for those of higher rank. Seventy-five percent of hired foreigners received salaries that were appropriate for upper civil service ranks. The majority of the foreigners were young, between twenty-six and thirty-five years of age.[26]

The upper echelon, around 7 percent of foreign employees, received salaries equivalent to those of department ministers, and 1 percent received in excess of double the prime minister's salary. A foreign engineer in the Railway Bureau commented that Japanese received one-sixth the pay of foreign employees. As a result, 5 percent of government expenditure in the Meiji period was used for foreign employees. This needs, however, to be compared to the 23 percent of the budget allotted during 1868 to 1886 for samurai stipends.[27]

Most of the professional foreign employees came from the four countries that were most important in Japan's foreign relations at the time: Great Britain, France, the United States, and Germany. Certain nationalities developed particular lines of work. In the late 1880s, the British took over from the French in naval training due to concerns regarding French technology. At the same time, the Japanese turned to German advisors for military advice, and they replaced the French in army training. And Americans established a foreign mail service at Yokohama.[28] The most significant American project was the technical assistance mission that went to the island of Hokkaido,[29] where activities included mining, railway construction, agricultural experimentation, and related industries.[30]

The largest group of British foreign workers was involved in public works engineering: railroads, telegraph lines, lighthouses, and harbor construction. In 1876 alone, more than one hundred British engineers and technicians were employed by the Ministry of Public Works to work on various projects, especially railway construction.[31] Although railway construction began in 1870, only seventy-six miles of railway line had been built by 1881. More important than the length of track, however, was the fact that by the end of the period Japanese technicians had been trained who could continue the work with little outside help, thanks to the founding of the Imperial College of Engineering by the Ministry of Public Works.[32]

Meiji leaders recognized that the employment of foreigners was a necessary—but temporary—evil. Thus they wanted to educate Japanese to replace foreigners as quickly as possible.[33] In response to this necessity, the Ministry of Public Works founded the Imperial College of Engineering in 1873. One of the British railway engineers, Henry Dyer (1848–1918), was appointed principal and served in this position until 1882. It took five years for the college buildings to be completed; the formal opening ceremony was held in 1878. The college consisted of six departments: civil engineering, mechanical engineering, telegraphy, architecture, mining, and applied chemistry and metallurgy. Among those who taught with Dyer was the British architect Josiah Conder, who would have a major role in designing Western-style public buildings and private homes throughout Tokyo. In 1885, when the Ministry of Public Works was abolished, the Ministry of Education took over responsibility for the college, which became the School of Engineering at Tokyo Imperial University.[34]

Technological change in Tokyo sometimes involved the adoption of select Western ideas to augment existing solutions, rather than the complete substitution of Western for Japanese know-how. This was certainly the case for the supply of water. We can turn to a paper titled "The Water Supply of Tokio," presented on 24 November 1877 by the Englishman Robert W. Atkinson (1850–1929), for an indication of how Tokyo (or Edo as it had previously been known) had a fairly adequate system already in place. Atkinson was a professor of analytical and applied chemistry at the University of Tokyo from 1874 to 1881. Prior to his appointment, he had been an assistant to A. W. Williamson, professor of chemistry at University College London.[35] In Japan, Atkinson made a name for himself with his research on *sake* and how to improve it.[36] On this occasion, however, the subject was water. Atkinson reported to the meeting of the Asiatic Society of Japan that, under his direction, three senior students had analyzed water samples, testing for the presence of organic matter as a sign of contamination by sewage. He concluded that wooden pipes should be replaced by others made of some impervious material. He also found that "Tôkiô water at its source is greatly superior to any of the London waters, but that it is inferior to the water supplied to Manchester and Glasgow."[37] Several years later, the German industrial chemist Oskar Korschelt, who taught at the Tokyo Medical School and at the school that later became the University of Tokyo from about 1875 to 1880, presented a follow-up paper on the city's water supply before the same society. In his paper of 12 December 1883, Korschelt argued that by digging sufficiently deep wells, pure water could be obtained everywhere in Tokyo: "To give to Tôkiô a really good water supply, it is only necessary to build reservoirs with filter beds at Yotsuya and Sekiguchi to clean the water if it has got turbid through rain, and replace the wooden pipes in the city by iron ones."[38] Korschelt found that the hollowed-out logs that transported the water provided an adequate water supply system, which could be improved relatively cheaply by replacement with metal pipes, deeper wells, and reservoirs.

With its fast-improving infrastructure, Tokyo came to be referred to also as *Teito* (Imperial Capital), especially from 1889, when the Meiji Constitution was promulgated. Various statues and monumental buildings were erected in public spaces, and construction commenced on the lavish Akasaka Detached Palace in an attempt to emulate the Palace of Versailles. Japanese planners looked to Paris as a model for the capital. They sought to construct wide boulevards and introduce a modern sewage system in what was referred to as the *Tōkyō no Parika* (Parisization of Tokyo).[39] But as the historian Henry D. Smith II suggests, "Tokyo in the late nineteenth century was less an 'imperial' city, home of the emperor, than an 'imperialist' city, seat of the empire."[40]

The Meiji-period slogan of *wakon yōsai* (Western science, Japanese culture) reflects how the Japanese saw foreign knowledge coexisting with Japanese culture and morals. By the late 1880s, however, the balance was becoming more Japanese, with foreign experts fast being replaced by Japanese who had studied abroad.[41] On 30 October 1890, the Imperial Rescript on Education was issued. In a sign of the times, it assured the Japanese that Japanese values were superior to Western technological knowledge. The Rescript signaled a shift from the national emphasis on practical learning to moral education: loyalty to and respect for the emperor.[42]

Western ideas and institutions were crucial in the establishment of a national identity in Japan, and foreigners were important agents in the transition to modernity. There were tensions between the need to rely on foreign experts and the desire of the Japanese to be independent. But in time, Japanese experts replaced the foreigners and a Japanese empire was soon in the making.

PART II REFORMULATING THE PAST IN JAPAN: LEARNING FROM THE WEST AND THE ESTABLISHMENT OF MUSEUMS IN TOKYO

As we have seen, a plan for the capital city of Tokyo was a crucial part of Japan's program of modernization during the Meiji period. Integral to this new vision for Japan was a reevaluation of the past in order to situate it with respect to the future envisioned by the modernizers. David Harvey has pointed out that at times of great change, a "desire for stable values leads to a heightened emphasis upon the authority of basic institutions— the family, religion, the state."[43] In Japan, the emperor and the past were key symbols in uniting the nation. Historian Stefan Tanaka has recently argued that the "discovery and separation of the past is one of the central components of the Meiji period."[44] We know much about how Japan was able to emancipate itself from its past by embracing a Western-inspired modernity, but little about the historical specificity of modernity in Japan.

This section will examine how the idea of modernity was diffused during the early part of the Meiji period, especially through the establishment of museums in Tokyo. The idea of the museum was foreign to Japan, and so we will explore how museums come to the attention of the Japanese, as well as what their aims were, what was included in their collections and exhibitions, and how they fit with the ideology of modernization. Elite groups of Japanese and Westerners were important in the reformulation of the past that took place through the museums. We will consider the roles of key Japanese who traveled overseas in the 1860s and the activities of the zoologist Edward S. Morse (1838–1925), who helped bring the Western theory of evolution to Japan.

SEEKING KNOWLEDGE FROM THE WEST

During the second half of the nineteenth century, Japanese statesmen and scholars, including Fukuzawa Yukichi (1835–1901), Kume Kunitake (1839–1931), and Sano Tsunetami (1822–1902), traveled abroad and returned to Japan with reports on the important role of the museum in linking knowledge to progress and national strength. The South Kensington Museum, in particular, resonated with the Meiji program of modernization, which involved the introduction of technology and the development of export industries. Although *bussankai*—exhibitions of rare foreign goods and plants, animals, and minerals from areas all over Japan—were held during the Tokugawa period, the idea of the museum was new to Japan. As we will see, the founding of the first museums played an important part of the creation of modernity in Meiji Japan.

Fukuzawa Yukichi came from a low-ranking samurai family in Osaka. He first went abroad in 1860, sailing to America on a Japanese ship along with three Japanese envoys who were traveling to Washington to ratify the Treaty of 1858. He also accompanied a delegation sent to Europe in 1862. This group visited France, England, Holland, Germany, Russia, Portugal, and England. During their six weeks in London, they attended the fourth International Exhibition at South Kensington. Fukuzawa and other members of the mission were impressed by the Western technology on display:

Manufactures from all countries and newly developed machines have been collected here to be shown to people … and in addition artisans have been sent to show how the machines are operated. Steam machines to make cloth from cotton and wool, chemicals to make ice in summer, steam-powered equipment for pumping water, all are here. There is fire-fighting apparatus, ingenious clocks, agricultural tools, horse trappings, kitchens, ship models, old books, and paintings without number.[45]

The trip formed the basis of Fukuzawa's famous book *Seiyō jijō* (*Conditions in the West*), which appeared in 1866. The book's great success established him as one of Japan's major authorities on the West.[46]

Shortly after Fukuzawa's book appeared, the Meiji Restoration of 1867–1868 ushered in a period of great change, during which Japan sought to emulate the West. In 1871, the Meiji government decided to send a special mission to the fifteen Western countries that had entered into treaties with Japan. Led by chief ambassador Iwakura Tomomi (1825–1883; figure 6.1), the Iwakura embassy (1871–1873) included the neo-Confucian scholar Kume Kunitake, whose role was to record the journey for the benefit of the Japanese people.[47]

FIGURE 6.1
Tokyo. *Prince Iwakura of Japan*, c. 1860–1875, wet-collodion process photograph. Brady-Handy Photograph Collection, Library of Congress Prints and Photographs Division, Washington, D.C.

Kume sought to understand how cities such as Chicago, New York, London, Paris, and Berlin had grown in population and prosperity—and generated so much technical innovation—over a relatively short period of time. Kume was particularly taken with museums, which he saw as serving an important educational role. In London, Kume visited the South Kensington Museum on 19 August 1872, where he was impressed by British workmanship and the links between the museum and industry. In addition, he felt that the Great Exhibition had engendered among the British a sense of pride in their own products.

Kume also visited the Smithsonian Institution, the British Museum, the Crystal Palace, and the Paris Conservatory of Arts and Sciences. He wrote:

When people go sightseeing in a museum, they see with their own eyes the orderly process by which a country becomes civilized. When a country flourishes, we learn from studying the basic reasons that this prosperity did not occur suddenly. There is always an order. Those who come first transmit their learning to those who arrive later. Those who awaken first inspire those who come awake later. By degrees there is advancement. We give this a name and call it progress. *By progress we do not mean tossing out the old and striving for the new.* A country develops by the accumulation of customs; it polishes the beauty of the past. ... There is nothing better than a museum to show this order.[48]

Through museums and expositions, Kume concluded, Western countries provided crucial links between past and future, and encouraged confidence, competition, and innovation.[49]

While the Iwakura embassy was visiting Europe in 1873, another Japanese delegation was on a study tour. This was the Ōkoku Hakurankai Jimukyoku (Austrian Exposition Bureau), which was responsible for Japan's participation in the international exposition in Vienna that year. In addition to promoting Japanese products and evaluating European products exhibited at the exposition, the group gathered information to help establish a museum in Japan. Their report was submitted in 1875 by Sano Tsunetami, the vice director of the Exposition Bureau and head of the mission.[50] The report recommended the founding of a new museum in Tokyo, modeled on the South Kensington Museum.[51] It would eventually become the Imperial Museum.

A strong advocate of museums was the American marine biologist Edward Sylvester Morse. He had originally come to visit Japan in 1877 to study brachiopods to test Darwinian theory. While there, he was invited to establish the Zoological Institute and a museum of natural history in the science department of Tokyo Imperial University.[52] Morse helped introduce scientific methods of collection and classification. One of his major achievements was the discovery and excavation of a prehistoric shell midden near Ōmori station in Tokyo. Many of the zoological and archaeological specimens that Morse and his students uncovered were displayed in the museum he had founded.[53] The more significant finds were exhibited at the Educational Museum in Tokyo, where Morse also was active from 1877 through 1879.

The discovery of the Ōmori mound called into question the ancient myths of the origin of the Japanese people and the idea that the imperial line had been founded by the Sun Goddess, for which there was no physical evidence.[54] Morse claimed, rather shockingly, that the pottery, tools, and other remains uncovered suggested that there was evidence that the Japanese, or at least those who had preceded them, had been cannibals.[55] The Japanese were required to rethink their history and to acknowledge a prehistory.

Morse was instrumental in opening the first biological station at Enoshima in 1877, and in 1878 helped found the first society of biology in Japan, the Biological Society of Tokyo University. More importantly, as the first professor of zoology at Tokyo Imperial University, Morse helped to introduce evolutionary theory to Japan. But for the Japanese, Darwinism was accepted as a theory of social as well as biological evolution.[56] It was evident to everyone throughout the world that Japan was very much a nation in evolution, as showcased by the museums and expositions. As Robert Kargon has argued in the cases of Western Europe and the United States, the Japanese also held "the conception of the inevitability of scientific and technical progress, leading towards a rewarding future."[57] And as Sophie Forgan has observed, "modernity was more often than not considered to be evolutionary rather than revolutionary in character."[58]

THE EDUCATIONAL MUSEUM

The Educational Museum was founded in Tokyo in 1872 as the Monbushō Hakubutsukan (literally, "Museum of the Education Ministry"), a pedagogical museum. Built to display the exhibits that had been gathered for the Vienna World Exposition in 1873, it provides a direct link between

Japanese participation in international expositions and the establishment of museums in Japan.[59] The museum underwent several name changes. In 1875, it was called the Tokyo Museum; in 1877 when it moved to a new building in Ueno Park, it was renamed the Educational Museum; and the name changed yet again in 1881 when it became known as the Tokyo Educational Museum.[60]

David Murray, an American, was important in building the museum's collection. He had arrived in Tokyo in 1873, on leave from his position as professor of mathematics and astronomy at Rutgers College. The Japanese government gave him the task of modernizing the nation's education system, and the museum was one way of showing the Japanese how it could be done. Speaking at the opening of the museum at its new location on 18 August 1877, he emphasized that the museum's purpose was to promote education in Japan, hence the emphasis of the book collection on pedagogy, textbooks, educational statistics, and reference. He took some pride in the collection of apparatus for teaching the physical sciences. As for natural history, Murray stated that "book-study alone is worse than useless. The student must have the real specimen to examine."[61]

Schoolroom equipment for kindergartens through higher schools was on display in the museum so that teachers could familiarize themselves with what was available. The displays also provided Japanese artisans with models for manufacturing what the teachers might require. Murray went so far as to say that the collection might "readily hereafter serve as an emporium for the collection and supply of material needed for equipping the schools of Japan."[62] Yet he emphasized that the museum was foremost an educational institution where visitors could learn by seeing what was on display. The museum would provide professors and their students with the means they needed to pursue research, and its technical staff would in turn train apprentices in Western methods.

The government actively encouraged the growth of the museum's collection, which benefited from Japanese participation in international expositions. This can be seen in David Murray's activities in the United States on behalf of the Japanese. In 1876, he contributed to the Japanese display on education at the Philadelphia Centennial Exhibition; he is listed in the *Official Catalogue of the Japanese Section* as an "agent of the Department of Public Education."[63] Murray also accompanied Japanese officials on a tour of educational facilities in the United States and used the exhibition as an opportunity to gather materials for the Educational Museum, which would promote science education in schools and colleges.[64]

In an 1877 report, Murray describes his efforts to collect materials for the museum, some of which were gifts. He hoped to make a collection "illustrating as far as possible the several branches of natural history, the arts and manufactures of western countries, and the appliances of education and civilization."[65] In this task, he sought the advice of a number of prominent people, including Joseph Henry and Spencer F. Baird (both of the Smithsonian Institution), Charles F. Chandler (School of Mines, Columbia College), J. D. Runkle (president of MIT), Edward S. Morse, John Eaton (U.S. Commissioner of Education), E. R. Beadle (mineralogist, Philadelphia), and George H. Cook (state geologist, New Jersey).[66]

American museums were anxious to acquire specimens of everything Japanese, from minerals, plants, and marine life, to stuffed animals, human skulls, and full skeletons. The museums were also keen to add examples of Japanese arts and crafts to their collections. By entering into exchange agreements with these institutions, the Educational Museum was able to build a collection at low cost. However, the agreements required the museum to collect large quantities of Japanese specimens in order to have something to trade. The Smithsonian made things easier by offering to serve as a distribution point for Japanese specimens destined for elsewhere in America.[67]

On 27 September 1877, the Ministry of Home Affairs ordered that archaeological artifacts excavated in Japan should be put on display at the Educational Museum. A similar order was issued soon after, in December 1877, by the vice minister of education, Tanaka Fujimaro (1845–1909), who requested that some of the more outstanding artifacts that had been found by Morse at Ōmori be exhibited at the Educational Museum.[68]

Morse visited the museum in 1877. In *Japan Day by Day* (1917), he describes how the halls of the main building were paneled in cedar and how much of the museum was devoted to the industrial arts, including models of coal mines, bridges, and dams. The models of bridges were particularly large and impressive, with some extending five or six feet in length. Morse also noted that collections of English ceramics had been sent from the South Kensington Museum. These were displayed in impressive display cases with French plate glass. In another building, many objects were on display that had originally come from the Vienna exposition, as part of an exchange, Morse surmised, for objects from the Japanese display. There were also rather mundane objects such as toothbrushes, pocketbooks, soap, penholders, and knives—things that, while familiar to Morse, may have struck the Japanese as being rather novel.[69]

He was impressed with the building, and the philosophy behind the collections:

The museum is a large, handsome, two-story building with a wing; a large library was in one of the lower halls and a long and spacious hall was filled with an extensive and interesting collection of educational apparatus from Europe and America—modern schoolhouses in miniature, desks, pictures, maps, models, globes, slates, blackboards, inkstands, and the minutest details of school appliances abroad. Despite the fact that every object was familiar to me, it was a most interesting museum and a kind that our larger cities at home should have. What a wise conception of the Japanese, entering as they were on our methods of education, that they should establish a museum to display the apparatus used in the work.[70]

Morse was less complimentary about the museum of natural history found on the second floor. He did, however, note the beautiful displays of fish.

Meanwhile, from the time of the Exhibition Bureau, the Ministry of Home Affairs had been developing its own plans for a museum of art and industry which would be housed in a new, purpose-built building in Ueno Park, designed by Josiah Conder. Before the museum's construction, the site was used for the First National Industrial Exhibition in 1877.[71] Morse visited the Industrial Exhibition on 28 August. He was particularly interested in a display of silk reeling, in which ten young kimono-clad women were seated on each side of a long table, reeling silk from cocoons.[72] "So few years have passed since the Restoration," Morse remarked, that he was "astonished, in going through the Exhibition, at the progress which has been made in the manufacture of objects which only a short time ago the Japanese were importing."[73]

The Ministry of Home Affairs' museum building was completed in 1880. Soon after, in 1881, administrative responsibility was transferred to the Ministry of Agriculture and Commerce, then to the Department of the Imperial Household in 1885, because the museum also took care of the imperial collection. It officially became the Imperial Museum in 1889.[74]

Morse left Japan in 1879, returning on 5 June 1882, after an absence of two years and eight months. During his absence a new, two-story

Zoological Museum had been constructed according to his plans.[75] Although Morse had high hopes for museums of natural history in Tokyo, without his presence, enthusiasm waned. University authorities lost interest in the museum and it was closed in 1885, when the science department moved to a new location at Hongo. The demise of the Educational Museum at Ueno Park soon followed in 1889, and the natural history section of the Museum of the Department of Agriculture and Commerce (the forerunner of the Imperial Museum) was also shuttered.[76] Interest in natural history had declined as a result of several factors: changes in the concept of what constituted science, the rise of laboratories, and the displacement of science exhibits from museums, which were increasingly devoted to art.[77]

In the 1870s and 1880s, museums taught by example. Morse relates in *Japan Day by Day* that the Department of Education held a meeting in Toyko of head teachers from throughout the nation to discuss their work. The teachers complained that schools could not afford the apparatus to teach the physical sciences. On hearing this, students from the Tokyo Normal School set about making various instruments for the study of physics before the teachers' eyes. While some of the instruments—balances, pendula, pumps, mirrors, camera obscura, and magnetic needles—were rather primitive, they nevertheless served the purpose and showed the ingenuity of the Japanese in being able to fashion apparatus from everyday materials.[78]

Speaking decades later about the connection between science, the city, and museums, Morse stated that "any improvements in city life must come through the efforts of the intelligent and thoughtful members of the community."[79] He argued that these elites could best appeal to the community by presenting a "scientific standpoint." He likened the city to a "huge organism," then proceeded to dissect and analyze it: "Above all, let us study its embryology and learn something of its origin and constituent parts. The great Agassiz once said a statesman should study natural history in order to fit himself properly for legislative work."[80] Morse also described a city's growth from the standpoint of a naturalist, starting with a pioneering family that lives alone in an isolated log cabin. A number of families arrive to form a village, which then grows into a town. The town eventually becomes a city. He contrasted the village with the city and all of its attractions: brightly lit streets, a public library, museums, concerts, lecture courses, theaters, etc. What was the reason for the difference? "If we ask ourselves the cause of this difference we shall find the answer in two words: communal effort."[81]

Morse made the case for public museums—and suggested what they should include in their collections—in an article that appeared in *Atlantic Monthly*:

First and foremost ... the town museum should illustrate the natural products of the immediate region, ... animals, plants, rocks, and minerals. ... Second, a general collection of similar material from elsewhere, to show the relation of the country to the rest of the world. ... And finally, a series of forms to show the phylogenetic development of the animal kingdom.[82]

In many respects, this was what had been achieved in Tokyo.

Museums helped Tokyo become Japan's center for Western learning, and the city itself became the face of Japan's Westernization. Museum buildings—which foreigners such as Josiah Conder helped design—were important elements of this public face. Conder taught architecture at the Imperial College of Engineering from 1876 to 1884. In his lectures to his students, he argued that "it is the permanence and beauty of its cities that contributes to the healthy growth of the empire and so it is your duty never to neglect your studies."[83]

As a result of the unequal treaties that Japan had signed with Western powers, the West, in effect, defined progress in Japan. Japan sought to become more like the West in the hope of revising the treaties, and Conder's task was to help accomplish this. He was commissioned by the Japanese government to design important Western-style buildings such as the Ueno Imperial Museum and the Rokumeikan, in which he combined Western architecture with some Indian Islamic design features. The former was a large, two-story, brick building completed in 1880. The Rokumeikan, or Deer Cry Pavilion, opened a few years later in 1883, was used to provide accommodation for foreign visitors and to entertain them at grand parties and balls. It has been described as an example of "English renaissance" architecture.[84] Whatever the label, it struck the Japanese as being exotic. Conder thought that a hybrid form of architectural style was appropriate for Japan, as it would bring together aspects of Eastern and Western culture, but the provenance of the style mystified many Japanese and he soon abandoned such Orientalist pastiches.[85]

Sometimes hybrid architecture was the result of compromises made because the Japanese architect hadn't been trained in the use of Western building materials such as bricks. In other cases, the hybrids grew out of a lack of Western materials: Japanese seeking to emulate a Western style would sometimes build a timber frame but face it with stone, stucco, tiles, or clapboards in what has been described as *giyōfu* (pseudo-Western style) or *wayō setchū* (Japanese-Western compromise) architecture.[86] These hybrid buildings, combining Japanese and foreign elements, could be seen in the foreign residential areas of Tokyo, Osaka, Yokohama, and Sapporo.[87]

PART III TOKYO: DISPLAYING THE FUTURE

In the late nineteenth century, national industrial exhibitions helped Tokyo to become an engine of change for the whole nation. We will next examine how Japan's new leaders borrowed from the West to articulate a new future by espousing an ideology of progress and emulating the international exhibitions on a domestic scale. These domestic industrial exhibitions expressed an ethos of progress and hopes for the future.

During the Meiji period many of the actors who facilitated these exhibitions had taken part in overseas missions and seen the West firsthand. They included the Meiji statesman Ōkubo Toshimichi, who was a former samurai from Satsuma, and Sano Tsunetami, the former samurai from Saga, who had been a member of the committee overseeing Japanese participation in Vienna's international exposition of 1873. Trusted foreign advisors such as Gottfried Wagener (1831–1892) also helped the Japanese to create a vision of the future that owed much to the West.

A major political and economic problem in Meiji Japan was the persistence of the unequal treaties that Japan had signed with Western nations from 1858. Because of these agreements, Japan lacked tariff autonomy and suffered the ignominy of extraterritoriality by other nations within Japanese borders.[88] As the Japanese strove to revise the unequal treaties, they were spurred into learning from the West, and as they did so, they sought to show the rest of the world that theirs was a civilized nation, worthy of respect. World's fairs and domestic industrial exhibitions provided the Japanese with opportunities to project a new face to the world, one that combined the old with the new.

We have seen previously how Iwakura Tomomi, the vice president of the Council of State, foreign minister, and court noble, was asked in October 1871 to head an overseas mission to countries that had signed treaties with Japan in order to study laws, customs, and institutions which could be adopted in Japan. The Japanese government also hoped that Iwakura's mission could persuade some of the countries to revise the terms of the treaties. Iwakura had four very capable deputies, two of whom were Ōkubo Toshimichi and Itō Hirobumi, the adopted son of a Chōshū samurai. They were accompanied by more than a hundred administrators, scholars, and students.[89] The Iwakura mission returned in 1873 with recommendations on technologies and institutions that should be introduced to Japan. As for revising the treaties, the mission was unsuccessful.

A full-scale attempt by the Japanese government in 1879 to end the unequal treaties was rebuffed by the British and French, while the United States and Germany showed themselves amenable to renegotiation. Thereafter the Japanese government made a deliberate effort to use the choice of foreign advisors and organizational models as a diplomatic weapon, by drawing on these two countries in preference to the previously favored British and French. Thanks to the efforts of Mutsu Munemitsu (1844–1897), the treaties were finally revised in 1894, with the renegotiated agreements going into effect in 1899, when extraterritoriality was finally abolished.[90] It would be several years before Japan secured tariff autonomy.

INTERNATIONAL EXHIBITIONS ABROAD
To ensure that their country would be perceived as respectable and modern—not just as part of an attempt to revise the unequal treaties, but to promote Japanese industry abroad—the Japanese put in great effort, including participation in international expositions.[91] Japan was first represented at the Paris World Exposition of 1867. After Japan was invited to participate by Napoleon III, Tokugawa Akitake, the younger brother of the shogun, was sent to France, accompanied by twenty-eight officials. The Tokugawa Shogunate and the provinces of Satsuma and Hizen mounted small-scale displays. Following the Meiji Restoration, the government ensured that a large Japanese display appeared at every major international exhibition until 1925, beginning with Japan's participation in the Vienna exposition of 1873.[92]

One of the aims of Japan's participation in world expositions was to build a strong and prosperous nation. With Japan increasingly on an expansionary footing, its government looked to the 1893 World's Columbian Exposition in Chicago as an opportunity to show that it was civilized and worthy of being considered a world power. Before the fair opened, the *Chicago Daily Tribune* commented, "Japan is the land of romance, and its recent efforts to be practical and business-like have not detracted from it. It is safe to say that the Japanese exhibit at Chicago will be one of the most interesting of the Exposition."[93] Although the Japanese exhibit did elicit public admiration, the *Chicago Daily Tribune* warned readers in 1894 that Japan's economic progress represented a potential threat. An article titled "There Is Danger in the East: Startling Industrial Development in the Japanese Empire" made this sentiment clear:

After the Paris Exposition of 1878 the Japanese Government imported from Europe the most improved machinery for cotton spinning and distributed it in certain provinces; in 1884 there were 35,000 bobbins; now there are over 380,000. ... With the best machinery of the West at their command, with the recognized artistic and inventive skill, with labor paid 12 cents a day for a man and six cents for a woman, they should be able to undersell any competitor.[94]

Such fears were understandable, given Japan's success in the Sino-Japanese War of 1894–1895. Yet, while some outsiders viewed Japan with caution, a decade later, its triumph in the Russo-Japanese War (1904–1905), combined with its success at the Louisiana Purchase Exposition in St. Louis in 1904, cast the country in a positive light. Despite the outbreak of the war shortly before the opening of the exposition (which celebrated the centennial of Thomas Jefferson's purchase of Louisiana from France), Japan still participated.[95] By this time, Japan had participated in more than two dozen such exhibitions.

The press response to Japan's participation in the Louisiana Purchase Exposition was positive. Japan, it was reported, had "the most extensive and picturesque site on a hillside, with dwarf pines, bridges, paths, and a variety of buildings."[96] The main Japanese exhibits were the Imperial Japanese Garden and a replica of the Golden Pavilion in Kyoto.[97] Four Ainu people were also on display, and drew comparisons with tribesmen from the Philippines, African pygmies, and "other uncivilized human specimens" also participating in exhibits at the St. Louis exposition.[98]

In contrast to the response of the American press, the Imperial Japanese Commission to the Louisiana Purchase Exposition compiled a publication, titled *Japan in the Beginning of the Twentieth Century*, to coincide with its participation there. The face which Japan sought to show Americans was that of a modern nation that excelled in agriculture, forestry, mining and metallurgy, fishing, and manufacturing, and which was keen to promote foreign trade. In addition, they wished to emphasize that Japan boasted a modern army and navy, and impressive communications and transportation networks. The commission highlighted how far Japan had come in terms of educational infrastructure and showed that it was becoming a colonial power in its own right in Formosa (Taiwan), which had been ceded to Japan in the aftermath of the Sino-Japanese War.[99]

The Louisiana Purchase Exposition Commission reported to President Theodore Roosevelt in 1906 that Japan's exhibits took up a total space "three times as large as that occupied by Japan at the Chicago world's fair of 1893 and the Paris Exposition of 1900, respectively."[100] Such extensive participation at international expositions involved great cost: the Japanese Diet had authorized total budgets of 630,000 yen for the Columbian Exposition in 1893, 1.3 million yen for the Paris Exposition of 1900, and 1.8 million yen for the Japan-British Exhibition in London in 1910. In addition to this were the costs absorbed by the prefectural governments and private firms that provided the bulk of the exhibits. Yet it was considered money well spent, for it provided opportunities by which Japan could absorb the latest innovations overseas and promote its own products.[101]

NATIONAL INDUSTRIAL EXHIBITIONS

The Japanese had long hoped to have the opportunity to hold a world exposition, but the high cost was prohibitive. Japan could afford to hold only domestic expositions, aimed more at promoting domestic industry than at displaying foreign products. Ōkubo, who was ultimately responsible for holding the First National Industrial Exposition, deliberately limited foreign exhibits to those purchased by the government. This is not to say that foreigners were not welcome to attend. Indeed, it was partly to impress foreigners with what Japan was capable of that the exhibition focused on the domestic rather than the international. For the benefit of diplomats, foreign employees, and members of the foreign press, exhibits displayed labels in English and catalogs in English were prepared.

A truly international exposition would have required the involvement and cooperation of foreigners in the selection and judgment of the products on display; but the Japanese were reluctant to invite official foreign participation. Gottfried Wagener did, however, serve as an advisor to both the 1877 and 1881 national industrial exhibitions, and wrote about the 1877 exhibition in *Meiji jūnen naikoku kangyō hakurankai hōkokusho* (Report of the 1877 National Industrial Exhibition).[102] Wagener had accompanied Sano to the Vienna World Exhibition in 1873, and was Japan's major Western advisor for the international exhibition held in Philadelphia in 1876. In both cases, Wagener was responsible for writing entries for the Japanese exhibits in the catalogs.[103] He served a useful role as a bridge between Japan and the West.

The staging of domestic industrial exhibitions, which became more ambitious and larger in scale over the years, served to promote Tokyo as the capital of Japan. The first three were held in Ueno Park. A new Exhibition Bureau, headed by Ōkubo, was established in 1876 within the Home Ministry to oversee the First National Industrial Exhibition. The bureau published *Naikoku kangyō hakurankai annai* (Guide to the National Industrial Exhibition) which introduced the layout of the pavilions and provided other information for visitors.[104] A major feature was a red brick building, especially constructed for the exhibition and later used as a museum.

The emperor and empress visited the exhibition on opening day, 21 August 1877. Their presence reminded all that the exhibits on display had been brought together from throughout Japan not only in the name of commerce but in service to the nation-state. The display of goods according to prefecture also reinforced this idea. As T. P. Keator, writing for the *Chicago Daily Tribune*, reported, "This is the first exhibition ever given under the auspices of the Government, and by Japan as a nation. Every province of the Empire is represented, and it is most interesting to note the first efforts of this, the latest nation to join the circle of civilized countries."[105] The event attracted over 450,000 visitors over the 102 days that it was open.[106] More than 16,000 exhibitors were spread over six different categories—mining and metallurgy, manufactures, art, machinery, horticulture, and agriculture—showing in seven pavilions. The machinery pavilion, arranged by the engineer Henry Dyer, showcased equipment made at a government factory that had ties with the Imperial College of Engineering, which Dyer had helped establish. The college, in turn, had strong links with the Ministry of Public Works headed by Itō Hirobumi.[107]

The award for the best invention, the *Hōmon Hōshō* (Firebird Prize), went to Gaun Tatchi (1842–1900) from the Shinshū region, for a "rattling spindle" (*garabō*) device for spinning yarn.[108] Cotton was packed into bamboo or metal tubes, drawn out from receptacles, and twisted onto rollers to become yarn.[109] Visitors certainly learned by seeing. The device was easy to copy and as there was no patent law at the time, little profit flowed to Gaun.[110] Gaun exhibited at the Second and Third National Industrial Exhibitions, held respectively in 1881 and 1890, winning prizes at all three.[111] However, due to competition from Western spinning machines, the popularity of the Gaun machine declined after 1890.

Tessa Morris-Suzuki has pointed out that most of the exhibits at the first exhibition drew on traditional agricultural and craft techniques. Though there were "modern" attractions, such as a steam locomotive, even in the machinery division exhibits tended to be made of wood and powered by waterwheels. Gaun's machine could be reproduced without too much difficulty by carpenters.[112]

The exhibitions displayed the old along with the new, in the hope of promoting new products that might draw on traditional technologies. This approach extended even to the fine arts. In this way, modernization in Japan was not simply a process of Westernization, nor was it simply a rejection of the past. The Ministry of Education in particular saw historical artifacts as "cultural assets that should be protected for the future."[113] As a reporter for the *Chicago Daily Tribune* wrote, exhibitions juxtaposed things Western with indigenous artifacts and technologies:

The exposition shows that the Japanese have changed, modified, revised, and, in many cases, have improved upon American and European productions. Some Japanese manufacturers have received a jaunty touch of Americanism, and some European and American productions have been enlivened with a genius Japanese. Sometimes the experimental combination is a success. Often it is a failure.[114]

It is not surprising that interaction occurred, with modification and adaptation occurring to meet Japanese needs and conditions,[115] for this reflected how the Japanese saw their cultural identity—as a hybrid sense of self which appeared outwardly Western (figure 6.2).

FIGURE 6.2
Tokyo. Detail of Toyohara Chikanobu, *Husband and Wife and Beauties on the Sumida River*, late nineteenth century, color woodblock print. Collection of Morris Low.

Attendance at the Second National Industrial Exhibition, in 1881, topped that at the first, with over 820,000 visitors over 122 days. The Ministry of Finance joined forces with the Home Ministry to hold the event. The centerpiece of the exhibition was the museum building designed by Josiah Conder (figure 6.3). The Western-style building underlined the authority of the new regime, serving to show that Japan was a modern nation, the equal of Western powers.[116] Another attraction—gas lighting—provided further evidence of Japan's modernity.

There is evidence that the expositions were indeed useful in promoting Japanese industry. For example, the Second National Industrial Exhibition provided craftsmen with an opportunity to showcase their products. The Tokyo-based Oki Kibatarō exhibited his Microsound device, an Edison-type telephone that used carbon powder instead of carbon rods. It won a second-place award for innovation.[117]

Handmade timepieces were also on display, including two wall clocks, a stand clock, and three pocket watches. However, the Japanese soon realized that such manually produced products could not compete with imported goods. In 1888, Hayashi Shihei established a factory to produce a thousand grandfather clocks a year using machine tools. Thus, the exposition provided the Japanese with an opportunity not only to showcase what they could do, but to confront the business realities of what competitors could offer.[118]

A major attraction at the Third National Industrial Exhibition, in 1890, was the electric trolley car, or tram, run by the Japan Electric Light Company. The company's name was emblazoned along the side of each passenger car for all to see.[119] A foreign visitor writing for the *Chicago Daily Tribune* proclaimed, "No city in the United States has a better system of electric lighting than Tokio. The exhibition of electric apparatus is superior and almost all of it is manufactured in Japan. … There is no patent on the electric apparatus in Japan, and electric lighting is consequently almost as cheap as kerosene."[120] Other important—though less spectacular—products included a photographic lens produced in Tokyo by Asakura Kametarō, the first such lens to be created in Japan, which won first prize.[121] Ueyama Eiichirō, from Wakayama prefecture, promoted the production of a new plant, pyrethrum, and won an award for merit. That year, he turned pyrethrum powder into mosquito-repellent coils. After considerable trial and error, he developed a marketable product by 1902.[122]

FIGURE 6.3
Tokyo. Detail of Utagawa Hiroshige, *Second National Industrial Exhibition at Ueno Park Showing the Art Museum and Fountain*, 1881, color woodblock print. Chadbourne Collection of Japanese Prints, Library of Congress Prints and Photographs Division, Washington, D.C.

Another success story was that of Mikimoto Kōkichi, who had established the first pearl farm in 1888 in Ago Bay and Toba. Oysters that had been implanted with foreign objects were placed in bamboo baskets and immersed in an attempt to produce pearls. Mikimoto exhibited natural pearls, pearl oysters, and pearl products at the Third National Industrial Exhibition. The zoologist Mitsukuri Kakichi (1858–1909) was one of the judges. Mikimoto's interest in cultured pearls came to the attention of Mitsukuri, who encouraged him and provided some useful advice. After many difficulties, Mikimoto finally succeeded in producing a semispherical pearl in 1893.

It was a fortuitous meeting. Mitsukuri had studied at Yale University (1877–1879) and Johns Hopkins University (1879–1881). In the summer of 1879, he had participated in the Chesapeake Bay Zoological Laboratory, an annual program of marine biology fieldwork in the oyster-rich waters of Chesapeake Bay, begun by Hopkins professor William Keith Brooks the previous year. The classes were a formative experience for Mitsukuri.[123] In 1881, Mitsukuri returned to Japan, and was appointed professor of zoology at Tokyo Imperial University the following year.[124]

From the late 1880s on, the Japanese government's vision of the future changed from one heavily reliant on Western institutions to one that looked to Japanese traditions for a guide. Since the national industrial exhibitions provided opportunities for the public to see and compare the latest inventions, different regions of Japan vied with each other for the best display. The exhibitions thus served, in varying degrees, to give citizens not only an idea of where Japan was headed but also where it had come from.

Meanwhile, many Japanese and foreigners had expressed dismay that the quality of Japanese arts and crafts had declined under pressure to produce for the export market. The *Times* of London reported that the 1877 and 1881 National Exhibitions had been largely commercial and industrial, and that "they bore evidence only too plain and painful of that sudden swift aberration from the canons of Japanese art taste, which attended the great changes begun in Japan some two-and-twenty years ago."[125] Partly to address this problem, old masterpieces of art from private collections were put on display at the Third National Industrial Exhibition, with some exhibits coming from the emperor's collection.[126] By promoting the idea of a shared culture dating back to antiquity, the government felt it could strengthen national identity and encourage new artistic and technological innovations.

Young Japanese such as Tokutomi Iichirō (1863–1957), known by his pen name, Sohō, and members of the group called the Min'yūsha (Friends of the Nation) had little patience for the past. Tokutomi's magazine *Kokumin no tomo* (The Nation's Friend) published an article on 22 June 1889, titled "What Value Has Foreigners' Flattery?" in which a critic wrote: "These foreigners regard Japan as the world's playground, a museum. ... They pay their admission and enter because there are so many strange, weird things to see. ... If our nation has become a spectacle, then we ought to be especially interested in the reform and progress that will make us a normal, civilized country."[127] Despite such criticism, the exhibitions did provide exciting spaces where visitors could experience modernity and achieve a type of visual literacy, within the safe confines of Ueno Park.[128] Although the Min'yūsha writers might not have been pleased, historical exhibits served to reassure some visitors that it was not by negating the past but by building on it that Japan could embrace the future. There were, arguably, no other such mass events in the Meiji era that were able to attract such large numbers of people to one location.[129]

In terms of lasting impact on the city, the exhibitions promoted the emergence of display-oriented retail spaces such as the *kankōba*, a type of bazaar in a building where retailers rented stalls. The first was established in 1878 by the Tokyo prefectural government in Eiraku-chō, Tokyo, in order to dispose of unsold exhibits left over from the First National Industrial Exhibition held the year before. By 1902, there were twenty-seven such bazaars to be found in Tokyo. By World War I, only two were left, as department stores had by then grown in popularity.[130]

During those years, Japan had emerged as a colonial power in its own right. In the same way that holding expositions was seen as part of the process of learning from the West, so too was the idea of imperial expansion. The Sino-Japanese War (1894–1895) was the result of conflict between Japan and China over primacy in Korea. Japan won, but was forced by France, Germany, and Russia to return the Liaodong Peninsula to China. Tensions between Russia and Japan over Korea led to an outbreak of war in 1904. Japan's victory in 1905 resulted in the Japanese gaining control of Korea. By 1910, Korea had been annexed by Japan as a formal colony.[131]

Rising levels of consumption and renewed confidence after the Russo-Japanese War led to calls for a Great Japan Exposition—an international exposition—to be held in 1912. As Iyenaga Toyokichi, lecturing at the University of Chicago in 1905 on the "Rise of Japan" suggested, "Civilization is not art, not commercial supremacy, not high tariffs or wages. ... It is the capacity to give society organic unity—to make the most of all its elements and members."[132] Expositions served this purpose. Arguing that "the greatest contemporary successes in national or social organization seem to be achieved under the pressure and stimulus of military aims,"[133] Iyenaga cited as evidence the prominence and success of Germany and Japan at the St. Louis exposition. Certainly in the case of Japan, this was true: between 1880 and 1912, almost one-third of government spending went toward the military, which also served to stimulate the economy.[134]

Despite Japan's good intentions, financial difficulties seem to have been the real cause of the postponement of plans for an international exposition. This was at a time when the military had to endure major cuts to their budgets. A truly international exhibition was still beyond what the Japanese could afford.[135] A sympathetic article in the *New York Times* on 3 September 1908 cited Robert Browning's famous statement that he judged people "by what they might be—not are, nor will be." The Japanese were praised for being ambitious, for aspiring to become a world power during the reign of the emperor Meiji.[136] Postponement until 1917, the writer felt, would enable the Japanese to celebrate the fiftieth anniversary of the Meiji period and the accession of the emperor. Eventually, however, such plans were dropped, not least because the emperor passed away in 1912.[137]

Ironically, the world's fair had been planned to emphasize that Japan's intentions were "industrial and peaceful, not warlike and aggressive."[138] The proposed site of the exposition, the Aoyama Parade Grounds, had been sold to the city of Tokyo by the national government, but it continued to be used as a drill ground for troops. With the outbreak of World War I in 1914, Japan joined the Allies in the conflict but was not engaged in much actual fighting.

The population of the city of Tokyo grew rapidly, almost doubling in the first two decades of the twentieth century. In 1900, the city boasted a population of 1.12 million; by 1920, the number had increased to 2.17 million.[139] As if to celebrate how far Tokyo had come, the Tokyo prefectural government hosted the Tokyo Peace Exposition in 1922. The venue was once again Ueno Park. The exhibition ran for 144 days, from 10 March through 31 July.¹⁴⁰ At a cost of some six million yen, the exhibition commemorated "world peace" and sought to showcase Japan's industrial progress since the end of World War I. The opening was attended by dignitaries including the honorary president of the exposition, General Prince Kan'in Kotohito, the adopted brother of the late emperor Meiji and a veteran of both the Sino-Japanese War and the Russo-Japanese War. Prime Minister Takahashi Korekiyo and Minister of Agriculture and Commerce Yamamoto Tatsuo were also in attendance, along with Tokyo mayor Gotō Shinpei. There was a decidedly colonial aspect to the exposition, with exhibits from Formosa, Korea, Manchuria, Mongolia, and Saghalien (Sakhalin) on display. These displays provided clear evidence of the growth of the Japanese empire, with Tokyo at its center.¹⁴¹

PART IV   REBUILDING TOKYO

The Great Earthquake of 1923 threatened all that Tokyo had achieved, as well as the city's future. The earthquake, centered at the head of Sagami Bay, which led into Tokyo Bay, hit Tokyo and Yokohama just before noon on 1 September 1923.¹⁴² It was Japan's worst natural disaster. The earthquake released energy equivalent to four hundred of the atomic bombs that were later dropped on Hiroshima. What was deadly was not so much the seismic force itself but the fires that broke out, as the quake struck while people were preparing their midday meals.¹⁴³ It is estimated that more than 140,000 people died, the same number believed to have died in Hiroshima. Ten thousand buildings, largely wooden, collapsed, and ultimately two-thirds of the city was leveled.¹⁴⁴

The disaster provided former Tokyo mayor Gotō Shinpei with an opportunity to articulate an ambitious new vision for Tokyo, though it would ultimately be thwarted. A physician and former colonial bureaucrat, Gotō was one of Japan's elite. He had learned from the West and sought to apply his ideas regarding urban planning and scientific colonialism to Japan and its growing empire. Gotō felt that it was incumbent on leaders such as him to furnish the city with the appropriate infrastructure. Like Edward Morse, Gotō saw cities as biological organisms.¹⁴⁵ As the chief civil administrator of Taiwan (1898–1906), he had advocated a "scientific" approach to colonial governance and development, an approach he referred to as "biological politics":

Any scheme of colonial administration, given the present advances in science, should be based on principles of Biology. What are these principles? They are to promote science and develop agriculture, industry, sanitation, education, communications, and the police force. If these are satisfactorily accomplished, we will be able to persevere in the struggle for survival and win the contest of "the survival of the fittest." Animals survive by overcoming heat and cold, and by enduring thirst and hunger. This is possible for them because they adapt to their environment. Thus, depending upon time and place, we too should adopt suitable measures and try to overcome the various difficulties that confront us. In our administration of Taiwan we will then be assured of a future of brilliance and glory.[146]

These ideas arguably reflect the influence of Gotō's early medical training and echo the views of the biologist Morse, discussed earlier, who saw the city as a "huge organism." These two men were not alone in their biological conception of the city. The Scottish biologist Patrick Geddes (1857–1932), author of *Cities in Evolution* (1915), wrote that the "intersocial struggle for existence" was no longer mainly dependent on the outcome of wars or industry: "Peace and prosperity depend above all upon our degree of civic efficiency, and upon the measure in which a higher phase of industrial civilisation may be attained in different regions and by their civic communities."[147] In this way, the evolution of cities was crucial to human progress.

Gotō had approached colonial development in Taiwan in a systematic and research-oriented manner that would have impressed Geddes. He had Taiwan's history and cultural traditions carefully documented in multivolume histories and biographical dictionaries. Huge compendia of fauna and flora of various places were also compiled.[148] At the same time, he made efforts to restructure the social and physical environment so that social change and evolution would occur. An effective infrastructure of schools, public health facilities, agricultural improvements, transportation and communications, and urban and port development was established to facilitate this.[149] The island was not unlike a laboratory in which experiments in social engineering were conducted.[150]

After his work in Taiwan, Gotō had sought to implement some of his ideas in Manchuria, where he was president of the South Manchuria Railway

Company from 1906 to 1908. Manchuria, like Taiwan before it, served as a laboratory for Japanese modernity. The railway was emblematic of the scientific and technological change that the Japanese hoped to usher in both at home and further afield in its growing empire.

Gotō served as mayor of Tokyo from late 1920 to 1923. In 1922, the year before the Great Earthquake, he invited Charles A. Beard, former director of the New York Bureau of Municipal Research, to Tokyo to help establish the Tokyo Institute for Municipal Research. In February of that year, Gotō wrote,

Although a city is not a nation, it is the nerve center of a nation. Therefore the rise and fall of a city unfailingly affects the rise and fall of a nation. This is why the Western nations are deeply concerned with the self-governing power of their cities and also with providing their citizens with a comfortable and secure life. At the same time, they are planning to organize a scientific city so that it can be a perfect utopia for mankind.[151]

Gotō's words were motivated by a desire to educate the residents of Tokyo on the need for "the operation of municipal government according to scientific methods."[152] He resigned as mayor on 25 April 1923, but remained committed to helping to improve Tokyo's administration with Beard's help. Beard wrote about how this could be achieved in his book *The Administration and Politics of Tokyo* (1923). In evaluating the organization and administrative methods of a municipal government, Beard suggested the need to first adopt a comparative approach, and study how various cities dealt with specific problems. Cities, he wrote, needed to encourage experimentation and be inventive when it came to problem solving. And finally, Beard argued that municipal governments could learn from the ways the private sector dealt with similar challenges.[153]

Following the Great Earthquake, press coverage indicated that like Gotō, some observers saw in the disaster an opportunity. Indeed, the day after the earthquake, Gotō joined the new cabinet as home minister. The lead article in the October issue of *The Far Eastern Review* published in Shanghai called this period "The Dawn of a New Era." The author, G. Bronson Rea, suggested that the calamity would turn out to be a blessing.[154]

As president of the newly created Imperial Capital Reconstruction Board, Gotō proposed that the central government purchase 33 million square meters of land in the burnt-out areas of Tokyo and spend 4 billion yen to create a rationally planned imperial capital. He urged the building of grand government buildings and wide boulevards, things planners had long hoped to create in Tokyo. A modern water supply and sewage systems, as well as public transport and a system of greenbelts and parks, were also part of his vision. Unfortunately for Gotō, the estimated cost of his plan was more than twice the entire national budget for 1923. In the aftermath of World War I, Japan could ill afford such grandiose plans.[155]

In early 1924, with the fall of the cabinet and Gotō's loss of the Home Ministry portfolio, the central government's Reconstruction Board became a Reconstruction Bureau. Although the Bureau wielded considerable influence on planning, it was the Tokyo municipal government that was forced to shoulder the bulk of the cost of reconstruction, some 300 million yen. The central government provided 147 million yen and paid the interest on a loan of 100 million yen from the Treasury. The funds were spent on road construction, improvements to waterways, and the construction of schools, bridges, and parks.[156]

What Gotō had been able to achieve in the colonial contexts of Taiwan and Manchuria proved more difficult to carry out in Tokyo after the Meiji period. A civil society was emerging, and opposition to his plans could not simply be quelled by military force. In Tokyo, citizens were able to voice their opinions, and politics were more fraught. As Beard later wrote, "The experience of London, San Francisco, and Tokyo raise the question whether any modern city can be planned except under a dictator! I do not refer to … grand boulevards such as Haussman cut in Paris or Burnham laid out in Chicago. I refer to city planning in the broadest sense."[157] A photograph album entitled *Fukkō* (Reconstruction) was published by the city of Tokyo in March 1930 to commemorate the capital's reconstruction. It opens with a photograph of the emperor in military dress, overlooking the city from an elevated position in Ueno Park, on 24 March 1930. This was the very location that the emperor had visited not long after the earthquake had struck, accompanied by officials—including Gotō. The message of the album was that a modern city had emerged triumphant, under the nominal gaze of the emperor. But the photograph album hid the reality of a Tokyo that had been rebuilt haphazardly, and even the emperor is said to have been disappointed that Gotō's full vision was not realized. Major trunk roads such as Shōwa Dōri and Hibiya Dōri in central Tokyo were reminders

FIGURE 6.4
Tokyo. *Edobashi Area, Shōwa Dōri*, from Tōkyō-shi, ed., *Fukkō* [Reconstruction] (Tokyo: Tōkyō-shi, 1930), hand-tipped photograph. Collection of Morris Low.

of what could have been achieved throughout the city if his vision had been adhered to (figure 6.4).[158]

Tokyo did emerge from the ashes, rebuilding along the lines of what had been there before—following the old street network. The streetcar tracks remained intact, as did much of the system of water and gas pipes. Landowners, big and small, were reluctant to lose their rights and were quick to rebuild their homes and businesses, not willing to wait for the possibility that their land might be claimed by the city for a new street or public park.[159] Gotō was a visionary, but he lacked the support not only of the citizens but of the leading political party, the Seiyūkai, which opposed his costly reconstruction plan.

The helter-skelter rebuilding of Tokyo can be viewed in a more positive light, however. As Henry D. Smith II puts it, "the city was no longer a 'showcase' for novelties from abroad, but [was] rather itself a powerhouse of innovation."[160] People and technology blended together, and the city evolved organically rather than being planned from above.

PART V   CONCLUSION

In the second half of the nineteenth century, a small modernizing elite, often of samurai background and based in Tokyo, learned from the West by employing foreign advisors and by going abroad. Tokyo provided them with a laboratory, a space in which to adopt and adapt Western-inspired concepts and institutions such as museums. Scientific and technological change was a key aspect of the changes that Japan underwent from 1868 through 1930. But the Japanese people needed to learn to accept the new institutions and foreign ways. The National Industrial Exhibitions encouraged the Japanese people to think in terms of the nation and to contribute to the national project of modernization. Tokyo thus became a symbol of progress and of Japanese hopes for the future (figure 6.5).

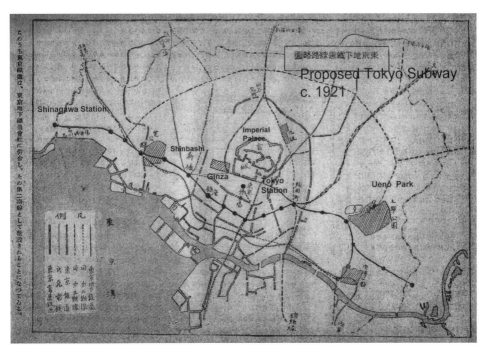

FIGURE 6.5
Tokyo. "Proposed Tokyo Subway Map," with added English, c. 1921, *Kagaku chishiki* [Scientific Knowledge], vol. 1 (1921), p. 17. Collection of Morris Low.

Tokyo was crucial to Japan's transformation during the Meiji period. Urban elites, especially the samurai and wealthy merchants, helped to construct a new Tokyo—and indeed a new Japan—through the establishment of science- and technology-related institutions, programs, and projects. The writings of Fukuzawa Yukichi were highly influential in promoting Western-style education and adapting ideas from the West. Various government slogans also helped to create an ideology of progress. Men such as Itō Hirobumi, Ōkubo Toshimichi, and Gotō Shinpei helped shape the character of Tokyo and the industrial growth in that city. Henry Dyer, Edward Morse, David Murray, Gottfried Wagener, and Josiah Conder were among the Western experts, generally young, who were relied on to provide expertise for a limited period to carry out specific projects when no Japanese experts were available. But the Japanese remained firmly in control. In time, the foreigners were replaced by Japanese, often those who had studied abroad. The modernity that Tokyo, and indeed Japan, embraced was a hybrid one, with heavy borrowings from the West. Western values and institutions were adopted and reformulated. Museums and exhibitions were especially important, functioning like two-way mirrors, which linked the past to Japan's future.

By the end of the nineteenth century, Tokyo became less of a laboratory for modernity, and visionaries such as Gotō in the meantime turned to the growing empire, with notable success in Taiwan and Manchuria. When he sought to emulate that success in Tokyo after the Great Earthquake, though, he met with resistance. Tokyo remained the center of Japan, but Gotō and other Japanese leaders looked to the growing empire with a sense of excitement.[161]

NOTES

1. All Japanese names in this paper are given in Japanese order, with family name first. In the notes, names are given as published. Sōseki's actual given name was Kinnosuke. He is customarily referred to by his pen name, Sōseki, rather than by his family name. Elongated vowels are indicated by macrons, except in well-known place names such as Tokyo (instead of Tōkyō).

2. Natsume Sōseki, *Sanshirō: A Novel*, trans. with a critical essay by Jay Rubin (Seattle: University of Washington Press, 1977), 17. For the Japanese-language text, see for example Natsume Sōseki, *Sanshirō* (Tokyo: Shūeisha, 1974).

3. Hisako Takahashi, "An Analysis of *Sanshirō* in Conjunction with the Visual Arts," *SOAS [School of Oriental and African Studies] Literary Review* 4 (Spring 2005): 1–20.

4. Ward William Biddle, "The Authenticity of Natsume Sōseki," *Monumenta Nipponica* 28, no. 4 (Winter 1973): 391–426.

5. Andrew Cornell, "Second Best and Suffering," *Australian Financial Review*, 7 June 2001, 60.

6. Hugh Clarke, "Sakutarō and the City," *Japanese Studies* 23, no. 2 (September 2003): 141–155.

7. Susan B. Hanley, "Urban Sanitation in Preindustrial Japan," *Journal of Interdisciplinary History* 18, no. 1 (Summer 1987): 1–26, esp. 3–4.

8. E. H. Norman, *Japan's Emergence as a Modern State: Political and Economic Problems of the Meiji Period* (New York: Institute of Pacific Relations, 1940), 11–12.

9. Akira Naito, *Edo, the City that Became Tokyo: An Illustrated History*, trans. H. Mack Horton (Tokyo: Kodansha International, 2003), 33–34.

10. Susan B. Hanley, *Everyday Things in Premodern Japan: The Hidden Legacy of Material Culture* (Berkeley: University of California Press, 1997), 105.

11. Ibid., 14.

12. W. G. Beasley, *The Meiji Restoration* (Stanford: Stanford University Press, 1972), 41.

13. Ibid., 108.

14. Ibid., 115–116.

15. Louis G. Perez, *Japan Comes of Age: Mutsu Munemitsu and the Revision of the Unequal Treaties* (Madison, NJ: Fairleigh Dickinson University Press, 1999), 188.

16. Beasley, *The Meiji Restoration*, 88–89, 96–97; Norman, *Japan's Emergence as a Modern State*, 38–40, 42–43.

17. Norman, *Japan's Emergence as a Modern State*, 49–50.

18. Beasley, *The Meiji Restoration*, 404.

19. Ibid., 325, 335.

20. D. Eleanor Westney, *Imitation and Innovation: The Transfer of Western Organizational Patterns to Meiji Japan* (Cambridge, MA: Harvard University Press, 1987).

21. Johannes Hirschmeier, "Shibusawa Eiichi: Industrial Pioneer," in William W. Lockwood, ed., *The State and Economic Enterprise in Japan: Essays in the Political*

Economy of Growth (1965; repr., Princeton: Princeton University Press, 1969), 209–247, esp. 209.

22. Gustav Ranis, "The Community-Centered Entrepreneur in Japanese Development," *Explorations in Entrepreneurial History* 8 (1955–1956): 80–98, esp. 92.

23. Hirschmeier, "Shibusawa Eiichi," 236.

24. Eikoh Shimao, "Some Aspects of Japanese Science, 1868–1945," *Annals of Science* 46 (1989): 69–91, esp. 72.

25. The Commission on the History of Science and Technology Policy, ed., *Historical Review of Japanese Science and Technology Policy* (Tokyo: Society of Non-Traditional Technology, 1989), 21–22.

26. Ardath W. Burks, "The West's Inreach: The *Oyatoi Gaikokujin*," in Burks, ed., *The Modernizers: Overseas Students, Foreign Employees, and Meiji Japan* (Boulder: Westview Press, 1985), 187–206, esp. 194–195.

27. Hazel J. Jones, "The Griffis Thesis and Meiji Policy toward Hired Foreigners," in Burks, *The Modernizers*, 219–253, esp. 231, 234.

28. Robert S. Schwantes, "Foreign Employees in the Development of Japan," in Burks, *The Modernizers*, 207–217, esp. 209.

29. Ibid., 214.

30. Ibid., 210–211.

31. Tessa Morris-Suzuki, *The Technological Transformation of Japan* (Cambridge: Cambridge University Press, 1994), 74.

32. Schwantes, "Foreign Employees in the Development of Japan," 210–211.

33. Jones, "The Griffis Thesis," 224.

34. Shimao, "Some Aspects of Japanese Science," 73, 75; Nobuhiro Miyoshi, *Henry Dyer: Pioneer of Engineering Education in Japan* (Folkestone, Kent: Global Oriental, 2004).

35. Y. Kikuchi, "Samurai Chemists, Charles Graham and Alexander William Williamson at University College London, 1863–1872," *Ambix* 56, no. 2 (July 2009): 115–137.

36. R. W. Atkinson, *The Chemistry of Sake Brewing*, Memoirs of the Science Department, Tokio Daigaku, no. 6 (Tokyo: Tokio Daigaku, 1881).

37. R. W. Atkinson, "The Water Supply of Tokio," *Transactions of the Asiatic Society of Japan* 6, part 1 (1877–1878): 87–98, esp. 96.

38. O. Korschelt, "The Water Supply of Tōkiō," *Transactions of the Asiatic Society of Japan* 12 (1885): 143–165, esp. 163.

39. Evelyn Schulz, "The Past in Tokyo's Future: Kōda Rohan's Thoughts on Urban Reform and the New Citizen in *Ikkoku no shuto* (One Nation's Capital)," in Nicolas Fiévé and Paul Waley, eds., *Japanese Capitals in Historical Perspective: Place, Power and Memory in Kyoto, Edo and Tokyo* (London: RoutledgeCurzon, 2003), 283–308, esp. 286.

40. Henry D. Smith II, "Tokyo and London: Comparative Conceptions of the City," in Albert M. Craig, ed., *Japan: A Comparative View* (Princeton: Princeton University Press, 1979), 49–99, esp. 70.

41. Yoshio Hara, "From Westernization to Japanization: The Replacement of Foreign Teachers by Japanese Who Studied Abroad," *Developing Economies* 15, no. 4 (December 1977): 440–461.

42. Toshio Nishi, *Unconditional Democracy: Education and Politics in Occupied Japan, 1945–1952* (Stanford: Hoover Institution Press, 1982), 17–18.

43. David Harvey, *The Condition of Postmodernity: An Enquiry into the Origins of Cultural Change* (Oxford: Basil Blackwell, 1989), 171.

44. Stefan Tanaka, *New Times in Modern Japan* (Princeton: Princeton University Press, 2004), 27.

45. See Fukuzawa Yukichi's journal "Seikō-ki" (1862) in Fukuzawa Yukichi, *Fukuzawa Yukichi zenshū* [Collected works of Fukuzawa Yukichi] (Tokyo: Iwanami Shoten, 1962), vol. 19, 27–28; quoted in W. G. Beasley, *Japan Encounters the Barbarian* (New Haven: Yale University Press, 1995), 82.

46. Carmen Blacker, *The Japanese Enlightenment: A Study of the Writings of Fukuzawa Yukichi* (Cambridge: Cambridge University Press, 1964), 6–7.

47. Marlene J. Mayo, "The Western Education of Kume Kunitake, 1871–6," *Monumenta Nipponica* 28, no. 1 (Spring 1973): 3–67, esp. 3, 5.

48. Kume Kunitake, *Tokumei zenken taishi bei-ō kairan jikki* [A true account of the tour in America and Europe of the special embassy], 5 vols. (Tokyo: Council of State, Records Division, 1878), vol. 2, 108; quoted in Mayo, "Western Education of Kume Kunitake," 47, with her italics.

49. Katie Grevdig, "Kume Kunitake: The Iwakura Embassy and Museums," *Wittenberg University East Asian Studies Journal* 30 (Spring 2005): 101–112, esp. 108.

50. Ōkoku Hakurankai Jimukyoku, *Ōkoku hakurankai hōkokusho* [Report on the Austrian Exposition] (Tokyo: Ōkoku Hakurankai Jimukyoku, 1875).

51. Alice Yu-Ting Tseng, "Art in Place: The Display of Japan at the Imperial Museums, 1872–1909" (Ph.D. diss., Harvard University, 2004), 30–32.

52. Lawrence W. Chisolm, *Fenollosa: The Far East and American Culture* (New Haven: Yale University Press, 1963), 30.

53. Edward S. Morse, *Japan Day by Day, 1877, 1878–79, 1882–83*, 2 vols. (Boston: Houghton Mifflin, 1917), 1: 145, 245; Peter Bleed, "Almost Archaeology: Early Archaeological Interest in Japan," in Richard J. Pearson, Gina Lee Barnes, and Karl L. Hutterer, eds., *Windows on the Japanese Past: Studies in Archaeology and Prehistory* (Ann Arbor: Center for Japanese Studies, University of Michigan, 1986), 57–67, esp. 65.

54. John S. Brownlee, *Japanese Historians and the National Myths, 1600–1945. The Age of the Gods and Emperor Jinmu* (Vancouver: University of British Columbia Press, 1997), 3–5, 89–91.

55. Tanaka, *New Times in Modern Japan*, 42; Edward S. Morse, "Shell Mounds of Ōmori," *Memoirs of the Science Department, University of Tokio, Japan* 1, part 1 (1879): 17–19.

56. Osamu Sakura, "Similarities and Varieties: A Brief Sketch on the Reception of Darwinism and Sociobiology in Japan," *Biology and Philosophy* 13 (1998): 341–357, esp. 343.

57. Robert Kargon, chapter 4 in this volume.

58. Sophie Forgan, chapter 3 in this volume.

59. Mimi Hall Yiengpruksawan, "Japanese Art History 2001: The State and Stakes of Research," *Art Bulletin* 83, no. 1 (March 2001): 105–122, esp. 113.

60. K. Aoki and T. Nakagawa, "A Review of the Development of Science Museums in Japan," *Japanese Studies in the History of Science*, no. 17 (1978): 1–12; Tseng, "Art in Place,"46.

61. David Murray, "Address at the Opening at Tokyo of the Educational and Scientific Museum" (1877), Manuscript Division, Library of Congress, Washington, DC.

62. Ibid.

63. The Imperial Japanese Commission to the International Exhibition at Philadelphia, *Official Catalogue of the Japanese Section, and Descriptive Notes on the Industry and Agriculture of Japan* (Philadelphia: Japanese Commission, 1876), 5–6.

64. Tadashi Kaneko, "Contributions of David Murray to the Modernization of School Administration in Japan," in Burks, *The Modernizers*, 301–321; W. I. Chamberlain, ed., *In Memoriam, David Murray, Ph.D., LL.D., Superintendent of*

Educational Affairs in the Empire of Japan, and Adviser to the Japanese Imperial Minister of Education, 1873–1879 (New York: Privately printed, 1915), 11.

65. David Murray, "Report upon Collections Made at the Philadelphia International Exhibition for an Educational and Scientific Museum at Tokyo, Japan" (1877), 3, Manuscript Division, Library of Congress, Washington, DC.

66. Murray, "Report upon Collections," 4. For the involvement of these men in the development of American science, see Robert V. Bruce, *The Launching of Modern American Science, 1846–1876* (New York: Knopf, 1987), esp. chapters 20, 21, and 25.

67. Murray, "Report upon Collections," 5.

68. Shiina Noritaka, *Nihon hakubutsukan hattatsu shi* [The history of the development of museums in Japan] (Tokyo: Yūzankaku, 1988), 121.

69. Morse, *Japan Day by Day*, 1: 149–150.

70. Ibid., 282–283.

71. Tseng, "Art in Place," 48, 50.

72. Morse, *Japan Day by Day*, 1: 250–251.

73. Ibid., 254.

74. Aoki and Nakagawa, "A Review of the Development of Science Museums," 1–12.

75. Morse, *Japan Day by Day*, 2: 211.

76. Isono Naohide, "Contributions of Edward S. Morse to Developing Young Japan," in Edward R. Beauchamp and Akira Iriye, eds., *Foreign Employees in Nineteenth-Century Japan* (Boulder: Westview Press, 1990), 193–212, esp. 199–200.

77. Forgan, "The South Kensington Museum," 1.

78. Morse, *Japan Day by Day*, 2: 430–433.

79. Edward S. Morse, "Can City Life Be Made Endurable?," address delivered at the annual commencement of the Worcester Polytechnic Institute, 21 June 1900, 1; reprinted from *Journal of the Polytechnic Institute* (November 1900), pages not given.

80. Ibid., 2.

81. Ibid., 3.

82. Edward S. Morse, "If Public Libraries, Why Not Public Museums," *Atlantic Monthly*, July 1893, 114; quoted in Dorothy G. Wayman, *Edward Sylvester Morse: A Biography* (Cambridge, MA: Harvard University Press, 1942), 371.

83. Josiah Conder, "What Is Architecture?" (1878), cited in Terunobu Fujimori, "Josiah Conder and Japan," trans. Gavin Frew, in Kusumi Kawanabe, Nobuo Aoki, Mitsuyuki Tago, Takeo Inada, and Aya Yuzuhana, eds., *Josiah Conder: A Victorian Architect in Japan*, exh. cat. (Tokyo: East Japan Railway Culture Foundation, 1997), 17–21, esp. 17.

84. Kimimasa Abe, "Meiji Architecture," in Naoteru Uyeno, ed., *Japanese Arts and Crafts in the Meiji Era*, Centennial Cultural Council Series, Japanese Culture in the Meiji Era, 8 (Tokyo: Pan-Pacific Press, 1958), 175–198, esp. 185.

85. Conder, "What Is Architecture?"

86. Toshio Watanabe, "Josiah Conder's Rokumeikan: Architecture and National Representation in Meiji Japan," *Art Journal* 55, no. 3 (Autumn 1996): 21–27, esp. 24.

87. Gregory Clancey, "Modernity and Carpenters: Daiku Technique and Meiji Technocracy," in Morris Low, ed., *Building a Modern Japan: Science, Technology, and Medicine in the Meiji Era and Beyond* (New York: Palgrave Macmillan, 2005), 183–206, esp. 188–191.

88. The "unequal treaties" prevented the Japanese from setting their own tariffs. As a result, there was concern that Japan might be flooded by foreign imports. In contrast, Western countries could impose high tariffs on Japanese goods that were exported. The system of extraterritoriality ensured that foreign nationals only answered to their own national law even when living in Japan, and thus were under the jurisdiction of foreign consuls. See Louis G. Perez, "Revision of the Unequal Treaties and Abolition of Extraterritoriality," in Helen Hardacre, with Adam L. Kern, ed., *New Directions in the Study of Meiji Japan* (Leiden: Brill, 1997), 320–334, esp. 322–323.

89. Yoshinori Amagai, "The Kobu Bijutsu Gakko and the Beginning of Design Education in Modern Japan," *Design Issues* 19, no. 2 (Spring 2003): 35–44, esp. 36; Li Narangoa, "Japan's Modernization: The Iwakura Mission to Scandinavia," *Kontur: Tidsskrift for Kulturstudier*, no. 2 (2001): 14–22.

90. Sandra T. W. Davis, "Treaty Revision, National Security, and Regional Cooperation: A Mintō Viewpoint," in Hilary Conroy, Sandra T. W. Davis, and Wayne Patterson, eds., *Japan in Transition: Thought and Action in the Meiji Era, 1868–1912* (Rutherford: Fairleigh Dickinson University Press, 1984), 151–173, esp. 152.

91. David Bromfield, "Japanese Representation at the 1867 Paris International Exhibition and the European Response to It," in Alan Rix and Ross Mouer, eds., *Japan's Impact on the World* (Canberra: Japanese Studies Association of Australia, 1984), 133–150, esp. 133.

92. Amagai, "The Kobu Bijutsu Gakko," esp. 36–37; Ōkoku Hakurankai Jimukyoku, *Ōkoku Hakurankai hōkokusho*.

93. "Japan's Exhibit at the Fair," *Chicago Daily Tribune*, 24 September 1892, 10.

94. "There Is Danger in the East: Startling Industrial Development in the Japanese Empire," *Chicago Daily Tribune*, 31 October 1894, 7.

95. "Have No Fear for Fair," *New York Times*, 5 August 1914, 6.

96. William Martin Aixen, "St Louis's Exposition Biggest Show on Earth," *New York Times*, 20 June 1904, 7.

97. Neil Harris, "All the World a Melting Pot? Japan at American Fairs, 1876–1904," in Akira Iriye, ed., *Mutual Images: Essays in American-Japanese Relations* (Cambridge, MA: Harvard University Press, 1975), 24–54, esp. 47–48.

98. "Primitive Races of Mankind and the Average American," *New York Times*, 7 May 1905, SM5.

99. Yamawaki Haruki, ed., *Japan in the Beginning of the Twentieth Century* (Tokyo: Imperial Japanese Commission to the Louisiana Purchase Exposition, 1904).

100. Louisiana Purchase Exposition Commission, *Final Report of the Louisiana Purchase Exposition Commission* (Washington, DC: Government Printing Office, 1906), 220.

101. Robert S. Schwantes, "Japan's Cultural Foreign Policies," in James William Morley, ed., *Japan's Foreign Policy, 1868–1941* (New York: Columbia University Press, 1974), 153–183, esp. 160–161.

102. Yoshida Mitsukuni, *Bankoku hakurankai: Gijutsu bunmei shi teki ni* [World expositions: From a history of technology and civilization point of view], rev. ed. (Tokyo: NHK Books, 1985).

103. Gunhild Avitabile, "Gottfried Wagener (1831–1892)," in *Meiji no takara, ronbun hen* [Treasures of Imperial Japan: Selected essays], Nasser D. Khalili Collection of Japanese Art, 1, ed. Oliver R. Impey (London: Kibo Foundation, 1995), 98–123, esp. 98, 102, 104.

104. Kuni Takeyuki, *Hakurankai no jidai: Meiji seifu no hakurankai seisaku* [The age of exhibitions: The exhibitions policy of the Meiji government] (Tokyo: Iwata Shoin, 2005), 63.

105. T. P. Keator, "Japan," *Chicago Daily Tribune*, 30 November 1877, 7.

106. Tokyo National Museum, "The Garden and Tea Houses," www.tnm.go.jp/en/guide/map/garden.html.

107. Amagai, "The Kobu Bijutsu Gakko," 42–43.

108. Also sometimes referred to as Tatsuchi, Tokimune, or Tatsumune Gaun. For further information, see Shōji Ishida, "Dai ikkai naikoku kangyō hakurankai shuppin: Gaun Tatchi no menbōki fukugenki no sekkei" [An exhibit at the First National Industrial Exhibition: The design of a reconstruction of a Tatchi Gaun cotton spinning machine], *Anjō-shi rekishi hakubutsukan kenkyū kiyō* [Research bulletin of the Anjō City Historical Museum], no. 2 (1995), online.

109. For an illustration of the machine, see Morris-Suzuki, *The Technological Transformation of Japan*, 90.

110. Sara Harris, "The Making of an IP Nation," *Japan Inc* (December 2002), online.

111. For details, see Kitano Susumu, *Gaun Tokimune to garabōki; Hatsumei no bunka isan; wa bōshi, wa nuno no nazo o saguru* [Gaun Tokimune and the Garabōki spinning machine: The cultural heritage of inventions, solving the puzzle of Japanese spun cotton and Japanese cloth] (Tokyo: Agune Gijutsu Sentaa, 1994); Miyashita Kazuo, *Gaun Tatchi: Garabōki 100 nen no ashiato o tazunete* [Gaun Tatchi: Tracing 100 years of the Garabōki spinning machine] (Matsumoto, Nagano: Kyōdo Shuppansha, 1993).

112. Morris-Suzuki, *The Technological Transformation of Japan*, 83, 90.

113. Kazuto Sawada, "Second National Industrial Exhibition," National Museum of Japanese History, Sakura City, Chiba, Japan, www.rekihaku.ac.jp/e-rekihaku/130/cover.html (accessed 15 September 2009).

114. R.R., "Japan's Big Exposition," *Chicago Daily Tribune*, 28 June 1890, 9.

115. Janet Hunter, "Institutional Revolution: The Case of Meiji Japan," in Magnus Blomström and Sumner La Croix, eds., *Institutional Change in Japan* (London: Routledge, 2006), part 1, chapter 2.

116. Watanabe, "Josiah Conder's Rokumeikan," 21–27, esp. 22–23.

117. Oki Printing Solutions, "Our Innovations: 1880s," okiprintingsolutions.com/1880.html (accessed 9 September 2009, cached, no longer available).

118. Hoshimi Uchida, "The Spread of Timepieces in the Meiji Period," *Japan Review* 14 (2002): 173–192, esp. 185.

119. Terashita Tsuyoshi, *Hakurankai kyōki* [Exposition memories] (Osaka: Ekisupuran, 1987), 237; illustrated in Yoshida, *Bankoku hakurankai*, 119.

120. R.R., "Japan's Big Exposition," 9.

121. Philbert Ono, "PhotoHistory 1868–1919," PhotoGuide Japan website, photoguide.jp/txt/PhotoHistory_1868-1919 (accessed 15 September 2009).

122. Masahiro Uemura, "Great People of Osaka: Developing and Promoting Insecticide Together with Pyrethrum, Eiichiro Ueyama," IBO English Osaka Business Update 4 (2004), IBO [International Business Organization of Osaka] website, ibo.or.jp/en/2004_4/ud03.html (accessed 15 September 2009).

123. Philip J. Pauly, "Summer Resort and Scientific Discipline: Woods Hole and the Structure of American Biology, 1882–1925," in Ronald Rainger, Keith R. Benson, and Jane Maienschein, eds., *The American Development of Biology* (Philadelphia: University of Pennsylvania Press, 1988), 121–150.

124. Isono Naohide, "Mitsukuri Kakichi, 1858–1909," in Kihara Hitoshi, Shinotoo Yoshito, and Isono Naohide, eds., *Kindai Nihon seibutsugakusha shōden* [Biographical sketches of biologists of modern Japan] (Tokyo: Hirakawa Shuppansha, 1988), 100–106.

125. "The Coming National Exhibition in Japan," *Times* (London), 15 March 1890, 4.

126. "Tokio's Forthcoming National Exposition," *Chicago Daily Tribune*, 1 April 1890, 8.

127. "What Value Has Foreigners' Flattery?," *Kokumin no tomo* [The nation's friend], 22 June 1889; quoted in Kenneth B. Pyle, *The New Generation in Meiji Japan: Problems of Cultural Identity, 1885–1895* (Stanford: Stanford University Press, 1969), 85.

128. Yoshimi Shunya, *Toshi no doramaturgii: Tōkyō sakariba no shakai shi* [Dramaturgy of the city: A social history of Tokyo and "sakariba"] (Tokyo: Kōbundō, 1987), 133.

129. Yoshimi Shunya, *Hakurankai no seijigaku* [The politics of expositions] (Tokyo: Chūō Kōron sha, 1992), 130.

130. Hatsuda Tōru, *Hyakkaten no tanjō* [The Birth of the Department Store] (Tokyo: Sanseidō, 1993).

131. Tessa Morris-Suzuki, "Japanese Nationalism from Meiji to 1937," in Colin Mackerras, ed., *Eastern Asia: An Introductory History*, 2nd ed. (South Melbourne: Addison Wesley Longman, 1995), 189–207, esp. 197–198.

132. Quoted in "Organizing a Nation," *Chicago Daily Tribune*, 20 August 1905, B4. See also "Lack of Golf and Bad Words Helps Jap Civilization," *Chicago Daily Tribune*, 18 August 1905, 7.

133. "Organizing a Nation," B4.

134. Morris-Suzuki, "Japanese Nationalism from Meiji to 1937," 198.

135. Robert A. C. Linsley, "Why the Tokio Exposition Was Postponed," *Harper's Weekly*, 24 October 1908, 27; "Japan Exposition in 1917," *New York Times*, 28 August 1908, 6.

136. "Japan's Position," *New York Times*, 3 September 1908, 6.

137. Takashi Hirano, "Retailing in Urban Japan, 1868–1945," *Urban History* 26, no. 3 (1999): 373–392, esp. 377–378; Edward Seidensticker, *Low City, High City* (New York: Alfred A. Knopf, 1983), 113–114; Schwantes, "Japan's Cultural Foreign Policies," 161–162.

138. "Organizing a Nation," B4.

139. André Sorensen, "Urban Planning and Civil Society in Japan: Japanese Urban Planning Development during the 'Taisho Democracy' Period (1905–31)," *Planning Perspectives* 16 (2001): 383–406.

140. Terashita, *Hakurankai kyōki*, 251–252.

141. "Tokio Exposition Opens," *New York Times*, 11 March 1922, p. 10.

142. For details of the earthquake, see Bruce Bolt, *Earthquakes and Geological Discovery* (New York: Scientific American Library, 1993), 18–22.

143. J. Charles Schencking, "The Great Kanto Earthquake and the Culture of Catastrophe and Reconstruction in 1920s Japan," *Journal of Japanese Studies* 34, no. 2 (2008): 295–331, esp. 299–300.

144. Gregory Clancey, *Earthquake Nation: The Cultural Politics of Japanese Seismicity, 1868–1930* (Berkeley: University of California Press, 2006), 220.

145. David G. Egler, "Pan-Asianism in Action and Reaction," in Harry Wray and Hilary Conroy, eds., *Japan Examined: Perspectives on Modern Japanese History* (Honolulu: University of Hawaii Press, 1983), 229–236, esp. 231.

146. Yusuke Tsurumi, *Gotō Shinpei den* [Biography of Gotō Shinpei], 4 vols. (Tokyo: Gotō Shinpei Hakudenki Hensankai, 1937), 2: 26–27. Cited in Ramon H. Myers, "Taiwan as an Imperial Colony of Japan: 1895–1945," *Journal of the Institute of Chinese Studies* 6 (1973): 425–451, esp. 435.

147. Patrick Geddes, *Cities in Evolution: An Introduction to the Town Planning Movement and to the Study of Civics* (1915; repr., New York: Howard Fertig, 1968), 393.

148. Mark R. Peattie, "Japanese Colonialism: Discarding the Stereotypes," in Wray and Conroy, *Japan Examined*, 208–213, esp. 211.

149. Mark R. Peattie, "The Japanese Colonial Empire, 1895–1945," in Peter Duus, ed., *The Cambridge History of Japan* (Cambridge: Cambridge University Press, 1988), 217–270, esp. 229–230.

150. Ibid., 238.

151. Gotō Shinpei, "Tōkyō shisei chōsakai kankei" [Material relating to the Tokyo Institute of Muncipal Research], microfilm reel 16-20, Gotō Shinpei papers, National Diet Library, Tokyo; quoted in Yukiko Hayase, "The Career of Gotō Shinpei: Japan's Statesman of Research, 1857–1929" (Ph.D. diss., Florida State University, 1974), 189.

152. Ibid.

153. Charles A. Beard, *The Administration and Politics of Tokyo: A Survey and Opinions* (New York: Macmillan, 1923), 20–22.

154. George Bronson Rea, "The Dawn of a New Era: A Calamity Turned into a Blessing," *Far Eastern Review* (Shanghai) 19, no. 10 (October 1923): 629–642.

155. Schencking, "The Great Kanto Earthquake," 314.

156. David Vance Tucker, "Building 'Our Manchukuo': Japanese City Planning, Architecture, and Nation-Building in Occupied Northeast China, 1931–1945" (Ph.D. diss., University of Iowa, 1999), 127–129.

157. Charles A. Beard, "Goto and the Rebuilding of Tokyo," *Our World* 5 (April 1924): 11–21, esp. 21.

158. Described in Tucker, "Building 'Our Manchukuo,'" 131.

159. Beard, "Goto and the Rebuilding of Tokyo."

160. Henry D. Smith II, "Tokyo as an Idea: An Exploration of Japanese Urban Thought until 1945," *Journal of Japanese Studies* 4, no. 1 (Winter 1978): 45–80, esp. 69.

161. The Japanese presence in Manchuria, Shanghai, and other parts of China in the 1930s offered opportunities for architects and planners to construct not only buildings but also whole districts and cities.

7 CODA

MIRIAM R. LEVIN

Let us pick up the example that begins this book—the Lumière brothers' filmed views of cities similar in infrastructure, though distinctive in local details—and consider these likenesses and differences again in light of the intervening chapters on Paris, London, Chicago, Berlin, and Tokyo. The comparison has been possible because these cities have shared a common historical purpose: In the never-ending game of international emulation and competition that marked the second industrial revolution, the science- and technology-based city was a means to—and measure of—a society's standing in the club of industrial nations. During these years, urban elites turned their cities into hubs for national and international development.

The striking similarities we find in the material appearance, institutions, and objectives of cities so geographically distant stem in part from the agreement of these cities' elites on the terms of the game. Historically, the decades between 1850 and 1930 were pregnant with opportunities for those possessing the will, resources, and authority to engage in urban rebuilding, to authorize expositions, and to establish museums. Leaders shared an appreciation of society-shaping synergies among these scientific and technologically defined activities that led to the invention of a new urban culture. These men were focused on using science and industrial innovations to create healthful and attractive environments, profitable communication, social order, and institutions for extending these benefits nationally and internationally. And they founded or reformed sets of institutions, sub-bureaucracies, and policies supporting their objectives, based in each case on a logical approach to gathering and applying knowledge.

Thus, the synergies among urban rebuilding, expositions, and museums together resulted from and helped concretize the explanatory constructs for change that elites promoted in each of these cities. These synergies also characterize the general consensus that existed among leaders about the primacy of urban development as means to industrial power and national greatness. In the case of all five cities, we have shown that urban rebuilding, expositions, and museums provided occasions for visiting delegations to observe how others did things, to assess the challenges, and to present one's own accomplishments—in some cases this led to international cooperation; in others, delegates returned home with information that pointed the way ahead.

Long ago, Thomas More, the author of *Utopia*, observed that when cities are designed by the same people, they end up all looking alike.[1] It is true that if we compare the Unter den Linden in Berlin, Pall Mall or the Strand in London, Michigan Avenue in Chicago, Omotesando Boulevard in Tokyo, and Paris's Champs-Élysées, we find designs not necessarily by the same people, but at least by people who knew about one another, with similar tastes, agendas, power, and access to funds, and the ability to work more or less rapidly. These characteristics aptly define the actors at the heart of our study, who nevertheless were quite varied and who arrived at their posts of power by a variety of means. In Chicago, the center of power and influence was an oligarchy, the group of businessmen belonging to the Commercial Club; in France, Germany, and Japan, it was the emperor and his circle; in Paris, it was a cadre of elected and appointed officials and professionalizing experts in a democratic republic. All these men came to power after revolutions, war, or civil unrest. By contrast, in London, the center of power was a collection of businessmen, entrepreneurs, and activist scientists and engineers, often operating independently, in a more stable liberal state. In Japan, the emperor came late to the game, and had to enlist the help of foreigners in modernizing Tokyo to catch up and keep up.

The dynamic of change, therefore, was approached somewhat differently in each of these cities. The Kaiser—Protestant, moralistic, and Gallophobic—sought to modernize his capital city without relinquishing tradition, and avoided emulating Paris by excluding theaters and an opera from his city center. Instead, museums took precedence in Berlin, along with research institutes. In Tokyo, traditional Japanese spaces and interests competed with modern, Western-influenced plans and uses. In Tokyo and Berlin, rather than massive world's fairs, smaller ones were hosted, with great effort expended on putting modern institutions, scientific and technological achievements, and urban improvements on view for visitors from abroad. In London, private interests mounted modestly sized expositions with highly focused international themes.

Moreover, differences in local traditions, economic and administrative arrangements, and technical training made for distinctions in the engineering, design, and control of sewer systems, rapid transit, street lighting, exhibitions, museums, and research institutes in each city. Distinctive sociopolitical styles shaped each variant of industrial urban culture being created. In Paris, three generations of *polytechniciens* gave coherence to much of the city's development during these years. Changes activated in Paris, London, Berlin, Chicago, and Tokyo represented a variety of solutions—via the expression of what Alfred Chandler has described as different managerial styles—to the biggest common problem in the age of industrial capitalism: uncontrolled change.[2] The paths to urban modernity might have curved a bit differently from place to place, but builders in all of these cities were committed to science and technology as the basis for building an international culture of change.

Although this culture originated in cities, its creators intended it to branch out and link up with other systems and places to form national and international circuitry. We have shown that this culture was expansive and transformative on two levels. First, it combined the urban dynamic with empire building. Energetically financed, organized, and elaborated communication and transportation systems originating in the cities brought colonial enterprises and native populations into the orbit of urban elites. These relationships were formalized through the participation of colonial enterprises in international expositions hosted in the cities. In their wake, these expositions left impressive, centrally situated buildings and collections of artifacts from colonies and conquered lands. Biologists, anthropologists, archaeologists, and art historians turned these buildings into museums where they systematically organized these specimens in evolutionary sequences that brought national progress to the foreground.

Second, the culture of change stimulated shifts in social geography, as well as an unending increase in population size and expansion of city borders. It also paved the way for the incorporation of new forms of transportation, new industries, and new sources of energy that promised more fluid and rapid circulation. Thus, the invention and implementation of this scientifically based culture with its emphasis on ordering change extended a process of urbanization that has yet to cease. These *villes tentaculaires* (octopus cities) posed the challenge of dealing with growth and innovation that outstripped what had already been constructed. In each of these cities, urban planners emerged with graphic plans for the future that incorporated visions of urban growth. Daniel Burnham's plan for Chicago, like those for 1850s Paris and post-earthquake Tokyo, emphasized integrating existing technologies—in this case of the 1890s—into ensembles of buildings, boulevards, and parks, without much interest in the industrial sectors of Chicago or the socioeconomic dynamics of the city. Alternatively, in Paris, the Musée Social measured, tracked, and channeled information on growth into plans for worker housing and for the extension of infrastructure improvements into working-class neighborhoods. Looking to prepare for the future rather than reacting to the present, the architect Eugène Hénard presented plans for the City of the Future to representatives from municipalities around the world attending the first international Town Planning Conference in London in 1910 (figure 7.1). Here he proposed that town planners consider how to integrate the automobile, the airplane, and the factory into the urban equation.[3]

FIGURE 7.1
Eugène Hénard, architect's drawing, from Eugène Hénard, "The Cities of the Future," in *Town Planning Conference, London, 10–15 October 1910, Transactions* (London: Royal Institute of British Architects, 1911).

FIGURE 7.2
Mexico City, street scene, photograph, c. 1911. Harris & Ewing Collection, Library of Congress Prints and Photographs Division, Washington, D.C.

These geographic and temporal features of this culture of change imply that the five cities examined here were not unique, but rather seminal. They are prime exemplars of a profound cultural shift that took place in urban centers around the globe. It would be useful to examine New York and Buenos Aires during this period, or Vienna, Moscow, Mexico City, St. Louis, and Shanghai in the 1920s and early 1930s, from the perspective we have proposed (figure 7.2).

What we have sketched out in this book is the inauguration by elites of a process of alteration that seemed unstoppable once begun. Whether "damned always to alter and never to be,"[4] or a blessing to humankind, these urban hubs became places marked by continuous change within and exporters of change elsewhere. With the aid of urban development, institution founding, and expositions and museums, science and technology have become so naturalized into our habitat that we now all live in cities of striking uniformity. The philosopher Jean Baudrillard describes the outcome of this process in the late twentieth century very well: "The cities of the world are [now] concentric, isomorphic, synchronic. Only one exists and you are always in the same one. It's the effect of their permanent revolution, their intense circulation, their instantaneous magnetism."[5]

NOTES

1. Sir Thomas More, *Utopia*, ed. George M. Logan and Robert M. Adams, 2nd ed. (Cambridge: Cambridge University Press, 2002), Book II, 44.

2. Robert Heilbroner, "Technological Determinism Revisited," in Donald MacKenzie and Judy Wajcman, eds., *The Social Shaping of Technology*, 2nd ed. (Buckingham: Open University Press, 1999), 74–75.

3. *Town Planning Conference, London, 10–15 October 1910, Transactions* (London: Royal Institute of British Architects, 1911).

4. Karl Scheffler, *Berlin, ein Stadtschicksal* (Berlin: Reiss, 1910), 266f.

5. Jean Baudrillard, *Cool Memories, 1980–1985*, trans. Chris Turner (London: Verso, 1990), p. 85.

INDEX

Page numbers in italics indicate illustrations.

Abel, Frederick, 95, 107–108
Adams, Henry, 151, 160
Adams, Henry Percy, 122n58
Ader, Clément, 47
AEG (firm), 7, 168
Akasaka Detached Palace, 213
Albert (prince consort), 78, 81
Albert Hall, 104
Alderson, Victor, 141–143
Alphand, Adolphe, 16, 27, *28,* 34, 39
Altes Museum, 181
American Exhibition (1887), 105, 108
Ancient Monuments Act, 93
Armour, Allison, 150
Armour, J. Ogden, 137, 141
Armour, Philip D., 140
Armour, Philip D., Jr., 141
Armour Institute of Technology (AIT), 6, 139–143, 147, 150
Armour Mission, 140
Armstrong, Henry, 99
Art Nouveau, 44
Asakura Kametarō, 231
Asiatic Society of Japan, 213
Association de Vieux Paris, 32
Association of Berlin Merchants and Industrialists. *See* Verein Berliner Kaufleute und Industrieller

Association of French Urbanists, 32
Atkinson, Robert W., 213
Automobile Club, 85
Ayer, Edward E., 145–150
Ayrton, William E., 88, 93, 99

Baedeker, Karl, 88, 103, 120n39
Baguio (Philippines), 154
Baird, Spencer F., 220
Balfour, A. J., 78
Bauakademie, 172
Baudrillard, Jean, 259
Bazalgette, Joseph, 123n61
Beadle, E. R., 220
Beard, Charles A., 8, 238–239
Bebel, August, 170
Bechmann, Georges, 36
Becker, Bernard, *Scientific London,* 81
Beijing, 207
Belgrand, F. E., 16
Benjamin, Walter, 14
Bennett, Edward, 155
Berlin, 6–7, 256–257. *See also specific institutions and events*
 adult education in, 7, 173, 182, 184–185, 187–188, 192–193
 chemical industries, 168
 Hobrecht Plan for, 175–176, 178

Berlin (cont.)
  industrial exhibitions in, 111, 188–196, *191*
  market halls, 173–174
  modernization of, 169–170, 173, 196
  municipal services and infrastructure, 168, 174–178, 191, 195, 196–197
  museums and monuments, 7, 169, 179–182, 184, 186
  population, 7, 167, 175, 207
  public health and hygiene, 168, 170, 173–178, 193, 195–197
  scientific culture, 168–170, 172–173, 177
  scientific popularization in, 7, 173
  urban renewal, 179
  Volkshochschulen, 173
Berlinische Zeitung, 169
Bernard, Claude, 59n9
Bernstein, Eduard, 196
Bertillon, Jacques, 47
Bibliothèque Nationale, 16
Bienvenüe, Fulgence, 27, 29, 41
Biological Society of Tokyo University, 218
Birmingham, 117n7
Bismarck, Otto von, 180
Board of Education (United Kingdom), 80
Bodemuseum, 181
Booth, Charles, 76
Borsig (firm), 174
Bourdais, Jules, 39, 43
Bramwell, Frederick, 82, 113
Brecht, Bertolt, 169
British Association for the Advancement of Science, 82
British Empire Exhibition (1924–1925), 106, 132n153
British Museum, 85, 102, 206, 217
Brooks, William Keith, 233
Browning, Robert, 235

Bryce, James, 160
Burnham, Daniel, 6, 93, 109, 137, 139, 143, 152–154, 158–159, 239. *See also Plan of Chicago*
Busse, Fred A., 157

Cahan, David, 172
Calabi, Donatella, 23, 32
Catholic Church (France), 14–15, 17, 20, 22, 24, 30
Centennial Exhibition (1876), 219, 228
Centralverband Deutscher Industrieller, 189
Chadwick, Edwin, 97, 174–175
Chambrun, Joseph-Dominique-Aldebert de Pineton, comte de, 30
Chandler, Alfred, 257
Chandler, Charles F., 220
Château d'Eau, 45
Chesapeake Bay Zoological Laboratory, 233
Chevalier, Michel, 16, 58n6
Chevreul, Michel-Eugène, 30, 64n41
Cheysson, Jean-Jacques-Émile, 16, 26, 31, 65n45
Album de statistique graphique, 31
Chicago, 5–6, 256–257. *See also specific institutions and events*
  civic associations, 137–138, 151
  class divisions, 135
  Department of Fire Protection Engineering, 142
  municipal services and infrastructure, 133–134, 141
  museums, 144–145
  pollution, 133–134
  population growth, 133
  public health and hygiene, 133–134
Chicago Auditorium Theater, 105
Chicago Commercial Club, 78
Chicago Company, 143, 147
Chicago Globe, 147

Chicago Manual Training School, 137, 139–140
Chicago Tribune, 147, 149, 226, 228–229, 231
"Cities of the Future" (Hénard), *258*
Citizen's Association, 135, 137
City and Guilds Central Institution, 95, 96, 99
City Livery Companies, 81–82, 96–97
City of Paris pavilion (Exposition Universelle, 1900), 50
Cleveland, Grover, 143
Cleveland (Ohio), 154–155
Coil and Current (Frith and Rawson), 87–88
Colonial and Indian Exhibition (1886), 104
Commercial Club, 6, 9–10, 137–139, 145, 147–148, 154–155, 172, 256
Compagnie Française Thomson-Houston, 38
Conder, Josiah, 212, 221, 223, 231, 242
Condillac, Étienne Bonnot de, 25
Condorcet, Jean-Antoine-Nicolas de Caritat, marquis de, 17, 25
  Sketch for a Historical Picture, 55
Conservatoire National des Arts et Métiers, 37, 38, 53, 101, 200n60, 217
Cook, George H., 220
Cooke, Ernest, 141
Cooper Union, 140
Crerar, John, 137
Crompton, Rookes Evelyn, 89, 123n68
Crystal Palace, 104
Crystal Palace Electrical Exhibition, 87
Curie, Marie, 37
Curie, Pierre, 37

Dai Ichi Ginkō, 211
Daly, César, 27

Davioud, Gabriel, 16, 39, 43
Davis, George R., 148
Debs, Eugene V., 135
De Forest, Lee, 142
Delano, Frederic A., 137, 154–155
de la Rue, William, 117n13
Department of Science and Art (United Kingdom), 80, 82, 100–102
Deutsches Museum (Munich), 181
Devonshire, Spencer Compton Cavendish, duke of, 105
Devonshire Commission, 83, 100
Diderot, Denis, 25
Disraeli, Benjamin, 101
Donnelly, John, 80, 82, 100, 102, 125n82
Drexel Institute, 140
Dubois-Reymond, Emil, 170, 172
Duruy, Victor, 16, 37, 59n9
Dyer, Henry, 212, 228–229, 242

Earls Court, 105
Eastman, S. C., 147
Eaton, H. W., 139
Eaton, John, 220
École d'Architecture, 50
École de Médecine, 37
École des Ponts et Chaussées, 26–27
École Municipale de Physique et Chimie, 38
École Normale Supérieure, 37
École Polytechnique (EPT), 15–16
École Supérieure d'Électricité, 38
Edinburgh, 111
Edison Company, 79
Edo. *See* Tokyo
Educational Museum (Tokyo), 8, 218–222
Eiffel, Gustave, 29, 43
Eiffel Tower, 26, 29, 40, 43, 45–47, 52
Einstein, Albert, 170

electrical engineering, 141–142. *See also* Institution of Electrical Engineers
Electrical Handbook of London, 88, 90–91
electricity. *See also* lighting, gas and electric
 at exhibitions and spectacles, 110, 111–112, 144, 153, 193
 versus gas interests, 135
 generating, 93–95, 132n158
 and industry, 92, 172
 infrastructure for, 76, 84, 89, *90–91,* 92, 94–95, 102, 110, 122n54, 141, 159
 technical standards for, 89, 114
 and tourism, 87–88
 and urban planning, 21–22, 24
Electric Palace (London), 112
Elektrotechnische Gesellschaft, 172
Elgin, James Bruce, earl of, 209
Ellsworth, James, 146
Engels, Friedrich, 160
ethnography, 30, 54–55, 73n109, 73n116, 226
Evans, John, 117n13
Exposition Internationale d'Électricité (1881), 36
Exposition Universelle (1867), 31, 39, 59n9, 225
Exposition Universelle (1878), 38–40, 43, 54, 226
Exposition Universelle (1889), 23, 31, 33, 36, 38–39, 43–46, 50, 54
Exposition Universelle (1900), 33, 36, 38–39, *40,* 41–42, *44,* 44–47, 50, 54, 227

Felisch, Bernhard, 190
Ferranti, Sebastian de, 123n68
Ferry, Jules, 22, 24, 26, 51, 53–54
Field, Marshall, 135, *136,* 137, 139, 143, 146–147

Field Museum, 6, 139, 143–150, 156
First National Industrial Exhibition (1877), 8, 221, 227–228, 231, 233
Fisheries Exhibition (1883), 104, 111
Fleming, J. A., 87
Flower, W. H., 99
Foerster, Wilhelm, 7, 172–173
Ford Motor Company, 75
Forgan, Sophie, 218
Fort Sheridan, 137
Foye, James, 141
France
 Ministry of Education and Fine Arts, 24, 37, 53
 Ministry of the Interior, 27
Franco-British Exhibition (1908), 105, 108, 111, 114
Franco-Prussian War, 4, 174
Frankland, Edward, 84, 99
Freeman, Clarence, 142
Frémy, Edmond, 64n41
French Exhibition (1890), 128n109
Friedrich Wilhelm IV, 181
Friedrich-Wilhelm University, 181
Fukuzawa Yukichi, 215–216, 242

Galerie des Machines, 33, 43
Galton, Douglas, 80–82, 97–98, 107
Gare d'Orsay, 41
Garnier, Charles, 16, 30, 32, 34, 50
Gassiot, J. P., 117n13
Gaun Tatchi, 229
Gavey, John, 120n40
Gay, Hannah, 122n59
Gay, John W., 122n59
Geddes, Patrick, 109, 237
Geikie, Alexander, 102
General Electric Company, 38, 153
General Post Office, 80, 85–89, 96, 103, 109, 113–114
General Strike (1926), 76
German Exhibition (1891), 128n109

INDEX

Germany, political unification of, 180, 185
Gewerbeakademie, 172
Giddens, Anthony, 8–9
Glasgow, 117n7
Glazebrook, Richard, 81, 89, 102, 109, 113
Goldberger, Julius, 189
Goldberger, Ludwig Max, 190
Goldberger, Max, 195
Goschler, Constantin, 175
Gotō Shinpei, 207, 236–240, 242
Grand Palais, 33, 42, 44
Graves, Edward, 120n40
Great Earthquake (1923), 8, 207, 236, 238, 242
Great Exhibition (1851), 9, 81, 99, 104, 217
Great Fire (Chicago, 1871), 135
Great Fire (London, 1666), 110
"Great Upheaval," 135
Group Plan Commission (Cleveland), 154
Guerin, Jules, 157
Gugerli, David, 169
Guillerme, André, 36
Guimard, Hector, 42
Gunsaulus, Frank W., 140–141, 147–148, 150
Guthrie, Frederick, 120n33, 121n52

habitations à bon marché, 31, 63n33
Hagiwara Sakutarō, 306
Hakodate, 208
Haldane, Richard Burdon, 113–114
Hampstead Central Electric Light Station, 87
Hamy, Ernest, 30, 54–55
Hardy, Léopold, 43
Harper, William Rainey, 137
Harris, Townsend, 208
Harvey, David, 214
Haupt, Herman, 142

Haussmann, Georges-Eugène, 4, 15–17, 21–22, 24, 26–27, 30, 34, 36, 41–42, 51, 55–56, 155, 175–176, 196, 239
Hayashi Shihei, 231
Haymarket riot (1886), 135, 137
Heaviside, Oliver, 87–88, 121n47, 121n52
Helmholtz, Anna von, 175–176
Helmholtz, Hermann von, 170, 172–173
Hénard, Eugène, 31–33, 42, 45, 49, 57, 155, 258
Henry, Joseph, 220
Higinbotham, Harlow, 137, 143, 147–148, 150, 152
history, use of, 30, 32, 44–45, 49–51, 54–55, 173, 179–180, 188
Hobrecht, Albrecht, 175, 178
Hobrecht, James, 175–178, 196
Hobsbawm, Eric, 169
Hogg, Quinton, 140
Holden, Charles, 122n58
Holmes, Charles John, *94*
Hooker, J. D., 100
Horniman, Frederick, 96
Hôtel de Ville, 32, 50
"Humanity Guided by Progress" (sculpture), 45
Humboldt, Alexander von, 173
Hunt, Bruce, 120n42
Hutchinson, Charles L., 137, 150
Huxley, T. H., 84, 99
hygiene. *See* public health and hygiene
Hygienic Department (U.S. Army), 97

Ieyasu Tokugawa, 207
Illustrated London, 102
Imperial College of Engineering. *See* School of Engineering, Tokyo Imperial University

Imperial College of Science and Technology (London), 95, 99, 102, 113
Imperial Institute (London), 95
Imperial Museum (Tokyo), 217, 221–223
Imperial Post Museum (Berlin), 181
Imperial Rescript on Education, 214
Institution of Electrical Engineers (IEE), 80, 85, 88–89, 92, 114–115
Insull, Samuel, 137
International Electrical Exhibition (1881), 131n142
International Electrotechnical Commission, 89
International Exhibition (1862), 215
International Health Exhibition (1884), 104, 107, 109, 111, 131n140
International Hygiene Congress (1852), 174
International Inventions Exhibition (1885), 104, 110
International Society of Urbanists, 49
Italian Exhibition (1888), 128n109
Itō Hirobumi, 211, 225, 229, 242
Iwakura Tomomi, 216, 216–217, 225
Iyenaga Toyokichi, 235

Janssen, Jules, 47–48
Japan. *See also specific institutions and events*
  colonialism, 227, 234, 236–237, 239, 253n161
  as economic competitor, 226
  educational system, 219
  foreign experts in, 211–214, 225, 242
  international image, 224, 226–227
  Meiji Constitution, 213
  Meiji Restoration, 207, 209–210, 216, 225
  Ministry of Education, 212, 229
  Ministry of Finance, 231

Ministry of Home Affairs, 210, 220–221, 228, 231
Ministry of Public Works, 211–212, 229
modernization, 209–211, 214
national identity, 214, 229, 233–234
population distribution in, 207
and technological innovation, 229, 231–232
Tokugawa period, 207–210
unequal treaties, 206, 209–210, 223–225, 248n88
wartime successes, 226, 234–235
Japan-British Exhibition (1910), *106*, 107–108, 114, 132n151, 227
Japan Electric Light Company, 231
Japan in the Beginning of the Twentieth Century, 227
Jeffrey, Edward T., 152
Jevons, W. S., 101–102
Johnson, Tom, 154

Kaiser-Friedrich-Museum, 181
Kaiser-Wilhelm-Gesellschaft (Kaiser-Wilhelm-Institutes), 7, 168, 172, 196
Kan'in Kotohito, 236
Kargon, Robert, 218
Keator, T. P., 228
Kelvin, William Thomson, Baron, 89
Kerr, Alfred, 192
King's College London, 114
Kiralfy, Imre, 105–106, 128n110, 132n152
Koch, Robert, 170, 173, 192–193
Korschelt, Oskar, 213
Köstering, Susanne, 187
Kühnemann, Fritz, 189–190
Kume Kunitake, 215–217
Kunstbibliothek (Berlin), 181
Kunstgewerbemuseum (Berlin), 181
Kyoto, 207–208

Labrouste, Ernest, 16
Laitko, Hubert, 172–173
Latham, Baldwin, 176
Law and Order League, 137
Le Corbusier, 57
Leiter, Levi, 137
Le Play, Frédéric, 16, 31, 42, 48, 59n9
Lessing, Julius, 189
Levin, Miriam, 102
Liebig, Justus, 178
lighting, gas and electric, 17, 36, 41–42, 45, 56, 84, 102, 111–112, 114, 135, 231
Lilienthal, Otto, 193
Lockroy, Édouard, 26, 42–43, 45, 49
Lockwood, Frank, 145
Lockyer, Norman, 83, 93, 100, 103, 113, 126n88
Lodge, Oliver, 87, 121n47
London, 10, 256–257. *See also specific institutions and events*
 architecture, 94–96
 economic upheavals in, 76
 elite class, defined, 77–83
 exhibitions, 77, 104–112, 115
 expansion of, 4–5
 historic preservation, 93, 110
 housing the poor, 108
 as imperial center, 4–5, 112, 115
 municipal services and infrastructure, 76, 78, 84, 89, 92–93, 102, 104, 107–108, 110, 115–116
 museums, 77–78, 80, 96–103
 polytechnic institutions, 95–96
 population, 12n1
 public health and hygiene, 97, 107–108, 115, 123n61
 urban planning, 93
London, City of, 110
London Chamber of Commerce, 107
London County Council, 78–79, 84, 89, 93, 96, 106–108, 113

London Town Planning Conference (1910), 93, 109, 130n130, 158, 258
London Town Planning Conference (1911), 32
Louisiana Purchase Exposition (1904), 89, 130n129, 226–227, 235
Lubbock, John, 78–80, 83–84, 93, 107, 110, 113
Luckhurst, Kenneth W., 189
Lumière brothers, 1, 255

MacKaye, Percy, 161
Manchester, 111, 133
Manchuria, 237–238, 242
Manila, 154–155
Marconi, Guglielmo, 87, 121n47
Märkisches Museum, 181, 184
Marx, Karl, 160
McClement, William T., 142
McCormick, Cyrus, 137
McKim, Charles, 154
McMillan, James, 154
Meiji Restoration, 7
Merchant's Club, 155
Metropolitan Board of Works, 93
Mexico City, *259*
Meyer, Wilhelm, 7, 173
Mikimoto Kōkichi, 233
Miller, Oskar von, 200n60
Milne-Edwards, Alphonse, 30, 64n41
Ministry of Education Museum. *See* Educational Museum
Min'yūsha, 234
Mitsukuri Kakichi, 233
Möbius, Kurt, 187
modernity, defined, 8–11, 21
Moltke, Helmuth von, 180
Monbushō Hakubutsuka. *See* Educational Museum
Moody, Walter D., 157–158
Moore, Charles, 155
More, Thomas, 256

Morris, William, 117n8
Morris-Suzuki, Tessa, 229
Morse, Edward S., 215, 217–218, 220–223, 236–237, 242
Moscow, 207
Mumford, Lewis, 130n130, 159
Murray, David, 219–220, 242
Musée Carnavalet, 17
Musée de la Ville de Paris, 53
Musée de l'Homme, 54
Musée des Arts Décoratifs, 53
Musée d'Ethnographie du Trocadéro, 17, 30, 53–54
Musée Social, 16–17, 23, 26, 30–33, 48–49, 96–97, 258
Muséum d'Histoire Naturelle, 53–54
Museum für Naturkunde. *See* Natural History Museum (Berlin)
Museumsinsel, 180–182, 186
Museum of Education (Tokyo), 8
Museum of Ethnology, 181
Museum of Oceanography, 181
Museum of Pathology, 181–188, *182*, 197
museums, educational role of, 149–150, 184, 186–187, 217, 222–223
Mutsu Munemitsu, 225

Nadaud, Martin, 36
Napoleon III, 4–5, 9, 15–17, 20–21, 25, 56, 58n6, 174, 225
National Educational Association, 150
Nationalgalerie, 181
National Physical Library, 80–81, 89, 109
Natsume Sōseki, 205–206
Natural History Museum (Berlin), 181–182, 186–188, 197
Natural History Museum (London), 99
Nead, Lynda, 129n124
Neues Museum, 181

New England Magazine, 152
New York Times, 149, 235
Nightingale, Florence, 97
Northampton Institute, 92
Norton, Charles, 155, 157
Notre Dame de Paris, 40

Office of Works (United Kingdom), 80, 85, 97
Oki Kibatarō, 231
Ōkoku Hakurankai Jimukyoku, 217
Ōkubo Toshimichi, 210–211, 224–225, 227–228, 242
Olmsted, Frederick Law, Jr., 154
Ōmori mound, 218, 220
Opéra, 16, 34
Orth, Johannes, 184
Osaka, 108, 207, 209
Ostrogorski, Moisei, 160
Otter, Chris, 114, 127n99
Owen, Richard, 99

Palais de Chaillot, 57
Palais de Justice, 16
Palais de l'Électricité, 33, 45
Palais de l'Industrie, 43
Palais du Trocadéro, 30, 39–40, *40*, 45, 54, 57
Paris. *See also specific institutions and events*
  and aesthetics of built environment, 14, 17, 20, 25, 42, 52
  as capital of Western civilization, 4, 52
  elite class, 9–10, 13–14, 22–23, 24, 25–27, 29–30, 33–34, 37, 43, 45, 51–52, 55–56
  expositions, 4–5, 14–16, 23–24, 38–39, 41–43, 51, 57, 111, 115
  following Franco-Prussian War, 4
  governmental bureaus in, 22–23
  institutions of higher education in, 29–30, 37–38, 59n9, 68n63

Métro, 17, 23, 29, 33, 37, 41–42, 49–51, 56–57, 62n22, 68n67
as model city, 155
Municipal Council, 17, 24, 41, 62n22
municipal services and infrastructure, 17, *18–19,* 26–27, 29, 33–34, 36–38, 41–42, 48–50, 56–57, 70n77, 108
museification of, 14–15, 17, 20–24, 25, 30, 33, 52–56
public health and hygiene, 17, 22, 36, 47, 59n9
and transformative technologies, 14, 22–23, 25, 27, 29, 33–34, 37–39, 44–48, 51–52, 56–57
and urban communications, 17, 20
working classes in, 20, 30–31, 48–50, 56, 63n33, 68n67
Paris Commune, 11, 20–21, 116n6
Parkes, Edmund Alexander, 97
Parkes Museum, 97–98, 115
Parks, H. W., 140
Parsons, Lawrence, 4th Earl of Rosse, 100
Pasteur, Louis, 37, 59n9
Patent Office Museum, 101
Pavillon d'Économie Sociale, 48
Peach, Charles Stanley, 94–95
Pergamonmuseum, 181
Perry, John, 88
Perry, Matthew, 206, 208
Peters, Wilhelm, 182, 186–187
Petit Palais, 42, 44
Philadelphia, 133. *See also* Centennial Exhibition
Phillips, Alfred, 142
Physikalische Gesellschaft, 172
Physikalisch-Technische Reichsanstalt, 7, 168, 172
Picard, Alfred, 27, 33, 42, 44, 46, 49
Pickstone, John, 50
Planck, Max, 170

Plan of Chicago, 6, 93, 109, 137–139, 155–159, *156,* 258
Platt, Harold, 135
Playfair, Lyon, 83, 102, 126n96
Poëte, Marcel, *La promenade de Paris au XVIIe siècle,* 31–32
Poubelle, Eugène, 26, 36
Pratt Institute, 140
Preece, William Henry, 5, 80, 82, 85, *86,* 87–88, 100, 103, 109, 113–114, 120n40, 121n52, 132n156
Proust, Antonin, 53
Prussia, Ministry of Cultural Affairs, 181, 186
Prussian Academy of Sciences, 7, 167
Public Health Act, 124n71
public health and hygiene, 17, 22, 36, 47, 59n9, 97, 107–108, 115, 123n61, 133–134, 168, 170, 173–178, 193, 195–197
Pullman, George, 108, 137, 139, 146–147
Pullman strike (1894), 135
Putnam, Frederick W., 147

railway strike (United States, 1877), 135
Rathenau, Emil, 195
Raymond, H. M., 142
Rea, G. Bronson, 238
Ream, Norman, 146
Redwood, Boverton, 119n29
Regent Street Polytechnic, 140
Reulecke, Jürgen, 189
Revue générale de l'architecture, 27
Ribbe, Wolfgang, 176
Rieger, Bernhard, 103
Roberts, Owen, 132n156
Rokumeikan, 223
Roosevelt, Theodore, 227
Roscoe, Henry, 126n88
Rosen, Christine, 134
Royal Academy of Art, 206
Royal Albert Hall, 40

Royal Arsenal (Woolwich), 103
Royal College of Science, 84, 99. *See also* Imperial College of Science and Technology; Royal School of Mines
Royal Dockyards, 103
Royal Institution, 81–82, 87, 92
Royal Mint, 103
Royal School of Mines, 95. *See also* Royal College of Science
Royal Society, 79, 81, 85, 87
Rubinstein, W. D., 117n12
Rucker, Arthur, 93
Runkle, J. D., 220
Ruskin, John, 117n8
Russo-Japanese War, 226, 234–235
Ryerson, Martin, 137, 148

Saint-Gaudens, Augustus, 154
St. Louis world's fair. *See* Louisiana Purchase Exposition
St. Peter's basilica (Rome), 40
Saint-Simon, Henri, 16, 23, 26
San Francisco, 154
Sanitary and Ship Canal, 134
Sanitary Institute of Great Britain, 98
Sano Tsunetami, 215, 217, 224, 228
Say, Léon, 26, 31
Scheffler, Karl, 167, 179
Schellbach, Karl, 172
Schering (firm), 7, 168
Schneer, Jonathan, 123n62
School of Engineering, Tokyo Imperial University, 212, 223, 228–229
Science Museum, 81, 102, 206. *See also* South Kensington Museum
Scott, J. W., 148
Scudamore, Frank, 120n40
Second National Industrial Exhibition (1881), 229, 231, 233
Second Regiment Armory, 137

Senckenbergmuseum, 181
Service des Monuments Historiques, 16
sewage systems, 7, 10
and agricultural interests, 176, 178
Berlin, 168, 174–178, 196–197
Chicago, 133–134, 137, 142
London, 76, 78, 97, 107, 110
Paris, 16, 26, 34, 36, 41–42, 48–50
Shaw, Albert, 76
Shibusawa Eiichi, 211
Shinpei, Gotō, 8
Siegessäule, 180, 185
Siegfried, Jules, 26, 31, 48
Siemens, Werner von, 173, 195
Siemens (firm), 7, 103, 168, 174
Siewert, Horst, 175
Simmel, Georg, 192, 194
Simon, Jules, 22, 26, 48
Sino-Japanese War, 226–227, 234
Skiff, Frederick J. V., 147
Slaby, Adolf, 195
Smith, Henry D., II, 240
Smithsonian Institution, 217
Société d'Anthropologie, 53
Société d'Ethnographie, 53
Société du Musée des Arts Décoratifs, 53
Société Française des Architectes, 66n50
Société Française des Habitations à Bon Marché, 48
Société Française des Urbanistes, 48–49
Society for Photographing Relics of Old London, 110
Society for the Protection of Ancient Buildings, 110
Society of Arts, 81, 96, 100, 107, 113
Society of Telegraph Engineers. *See* Institution of Electrical Engineers
Sohō. *See* Tokutomi Iichirō

Sousa, John Philip, 153
South Kensington Museum, 80, 85, 99–103, 200n60, 206, 217, 220. *See also* Science Museum; Victoria and Albert Museum
Spottiswoode, William, 117n13, 125n85
Sprague, A. A., 137
Stead, William T., 162n14
Stine, Wilber M., 141–142
Strohmeyer, Klaus, 176
Strutt, John, 78–79
Sturges, George, 147
Sullivan, Louis, 151–152
Sutcliffe, Anthony, 196
Swan Electric Company, 79
Swinton, A. A. Campbell, 87

Taine, Hippolyte, 55
Taiwan, 227, 236–239, 242
Takahashi Korekiyo, 236
Tanaka Fujimaro, 220
Tanaka, Stefan, 214
Taylor, Fitzhugh, 142
Technical University of Berlin, 7, 167, 172, 196
Technische Hochschule (Charlottenburg), 140
techno-nostalgia, 153, 158
Telegraph Instrument Galleries, 88
telegraphy, 87–88, 92, 103
Third National Industrial Exhibition (1890), 229, 231, 233
Third Republic, and cultural democratization, 15, 17, 20–21, 23, 25, 34, 37, 39, 42, 51, 56
Thompson, Sylvanus P., 88, 92
Tiede, August, 187
time, constructed, 21, 42, 51, 53, 61n13
Tocqueville, Alexis de, 136–137
Tokugawa Akitake, 225

Tokugawa family, 7
Tokutomi Iichirō, 234
Tokyo, 7–8, 108. *See also* Japan; *specific institutions and events*
architecture and urban space, 212–213, 223–224, 231, 234–235, 256
hygiene, 208
international exhibitions, participation in, 206
modernization of, 8
municipal services and infrastructure, 213–214, 239–240, *241*
museums, 206, 215, 223
national industrial exhibitions, 206, 221, 227–229, 231, 233–234, 241
population, 207–208, 235
rebuilding of, 236–240
Tokyo Imperial University, 218, 222. *See also* School of Engineering, Tokyo Imperial University
Tokyo Institute for Municipal Research, 238
Tokyo Peace Exposition (1922), 235–236
Toyohara Chikanobu, *Husband and Wife*, 230
Trumbull, Morris, 143
Turner, Frank M., 117n15
Twain, Mark, 179
Tyndall, John, 100

Ueyama Eiichirō, 231
Union Central des Beaux-Arts Appliqués à l'Industrie, 53
University College, London, 81, 97–98. *See also* University of London
University of Chicago, 140, 143
University of London, 79, 113–114. *See also* University College, London
Unwin, Raymond, 130n130
Urania, 173, 200n60
urbanism, 31, 49

urban planning and urbanization, 2–3, 33, 109, 158, 176
Utagawa Hiroshige, *Second National Industrial Exhibition*, 232

Verein Berliner Kaufleute und Industrieller, 189–190
Verne, Jules, *Paris in the Twentieth Century*, 13, 49
Victoria (queen), 93
Victoria and Albert Museum, 99, 206. *See also* South Kensington Museum
Vienna, 207
Vienna World Exhibition (1873), 217–218, 220, 225, 228
Viollet-le-Duc, Eugène-Emmanuel, 16
Virchow, Rudolf, 170, *171*, 172–173, 175–178, 182–186, 188, 192, 195, 197
Volkshochschulen, 173

Wacker, Charles, 157
Wacker's Manual of the Plan of Chicago, 158
Wagener, Gottfried, 224, 228, 242
Wallace Collection, 206
Washington, D.C., plan for, 154–155
Webb, Aston, 95
Webb, Beatrice, 120n34
Webb, Sidney, 96, 120n34
Weber, Max, 24
Wellcome Chemical Research Laboratories, 130n131
Wells, H. G., 98, 101, 103
Wembley Stadium, 105–106, 128n111, 128n113
Westminster, duke of, 94
White City (London), 75, 105–107, 112, 115, 128n111
White City (World's Columbian Exposition), 75, 139, 144, 151–154
Wiebe, Eduard, 175–176

Wiebe, Robert, 6, 158
Wilhelm I, 180
Wilhelm II, 180, 189–190, 195, 196, 200n60
Williamson, A. W., 213
Wolfe-Barry, John, 119n29
Wood, Henry Trueman, 81–84, 100, 103, 107–108, 113
Woodcroft, Bennet, 101
World's Columbian Exposition (1893), 6, 75, 104–105, 134–135, 137–138, 139, 141, 143–144, *145*, 146–154, 226–227

X Club, 5, 79

Yamao Yōzō, 211
Yokohama, 212, 236

zaibatsu, 210
Zeublin, Charles, 153
Zobeltitz, Feodor von, 170
Zola, Émile, 23, 45, 51

HT 361 .U7173 2010

Urban modernity

GAYLORD